AF702685

Transforming Urban Transport

Transforming Urban Transport confronts head-on the dilemma faced by a world wedded to mobility: the danger of continuing along the fossil-fuelled path and the real paucity of viable technological alternatives that can be deployed in time. To respond to the dilemma, the ideal of urban transport must be changed from auto-based mobility to systems of sustainable transport in which public transport and non-motorized transport work together to reduce climate change pressures, enhance urban quality and preserve life and health.

The book challenges the commonly held view that a combination of urbanity and higher residential density expressed in compact cities – expected to have greater public transport use – will resolve urban transport/environment problems, instead showing that transport systems can be changed to meet the environmental imperatives without the massive spatial change implied in that commonly cited approach.

But the problem of change of urban transport is profoundly institutional and cultural. Changes in urban mobility and transport require local institutional policy action. To support such action, the book explores new methods of governance of transport in dispersed and concentrated cities, new techniques for assessing transport need, ways of improving childhood mobility, guidelines for political mobilization, and norms of knowledge sharing.

Drawing together leading scholars from different disciplines in Australia, Japan and China, this book provides a unique fusion of perspectives and engages with the coming needs of transport planning practitioners in both high density and dispersed cities.

Nicholas Low is Professor in Urban and Environmental Planning at the University of Melbourne, Australia.

Transforming Urban Transport

The ethics, politics and practices of sustainable mobility

Edited by Nicholas Low

Routledge
Taylor & Francis Group
LONDON AND NEW YORK

First published 2013
by Routledge
2 Park Square, Milton Park, Abingdon, Oxon, OX14 4RN

Simultaneously published in the USA and Canada
by Routledge
711 Third Avenue, New York, NY 10017

Routledge is an imprint of the Taylor & Francis Group, an informa business

© 2013 Nicholas Low selection and editorial material; individual chapters, the contributors

The right of Nicholas Low to be identified as author of the editorial material, and of the individual authors as authors of their contributions, has been asserted in accordance with sections 77 and 78 of the Copyright, Designs and Patents Act 1988.

All rights reserved. The purchase of this copyright material confers the right on the purchasing institution to photocopy pages which bear the photocopy icon and copyright line at the bottom of the page. No part of this book may be reprinted or reproduced or utilised in any form or by any electronic, mechanical, or other means, now known or hereafter invented, including photocopying and recording, or in any information storage or retrieval system, without permission in writing from the publishers.

Trademark notice: Product or corporate names may be trademarks or registered trademarks, and are used only for identification and explanation without intent to infringe.

British Library Cataloguing in Publication Data
A catalogue record for this book is available from the British Library

Library of Congress Cataloging-in-Publication Data
Transforming urban transport : the ethics, politics, and practices of sustainable mobility / edited by Nicholas Low.
p. cm.
Includes bibliographical references and index.
1. Urban transportation--Government policy. 2. Urban transportation--Environmental aspects. 3. Sustainable engineering.
I. Low, Nicholas.
TA1205.T68 2013 388.4028'6--dc23
2012019166

ISBN13: 978–0–415–52903–7 (hbk)
ISBN13: 978–0–203–08386–4 (ebk)

Typeset in Sabon
by Bookcraft Ltd, Stroud, Gloucestershire

To Elizabeth Low

Contents

Figures, tables and boxes

Figures

Tables

Boxes

Contributors

Courtney Babb is a PhD candidate in Urban and Regional Planning at Curtin University, Western Australia. His research focuses on the social impacts of transport planning and the way planners evaluate the quality of urban environments.

Matthew Burke is a Senior Research Fellow in Griffith University's Urban Research Program. His research focuses on travel behaviour, transport accessibility, urban environments and the relationships between active transport, physical activity and health.

Carey Curtis is Professor in City Planning and Transport at Curtin University, Australia. She is a Partner of the Australasian Centre for the Governance and Management of Urban Transport (GAMUT) and chairs the International Editorial Board of the journal *Urban Policy and Research*.

Leigh Glover is Director of the Australasian Centre for the Governance and Management of Urban Transport (GAMUT) at the University of Melbourne, Australia, where he is a Senior Research Fellow. His recent research concerns sustainable transport and institutional responses to sustainability issues, with a focus on the Asia-Pacific region.

Sun Sheng Han is a Professor in Urban Planning at the University of Melbourne. He is the Program Coordinator of the Master of Urban Planning, and a member of the editorial boards of *Progress in Planning*, *GeoJournal*, *Urban Planning International*, and the *Journal of Urban and Regional Planning*.

Damon Honnery is Associate Professor in Mechanical and Aerospace Engineering, and Deputy Director of the Laboratory for Turbulence Research in Aerospace and Combustion at Monash University, Melbourne. He is a Lead Author for Working Group 3 of the IPCC 5th Assessment Report. He is also an Honorary GAMUT Fellow.

Crystal Legacy is a Research Associate at the City Futures Research Centre at the University of New South Wales, Australia. Crystal completed her PhD from the GAMUT Research Centre in 2010. She is the co-editor of the book *Building Inclusive Cities: Women's Safety and the Right to the City* (Routledge, 2012).

Kevin O'Connor is Professorial Fellow in Urban Planning at the University of Melbourne and at the School of Geography and Environmental Science at Monash University. He is also a GAMUT Partner. His current research is on the impact of logistics activity on cities, with an emphasis upon the development of airports. He is currently an Associate Editor (Asia-Pacific) of the Journal of Transport Geography.

Nicholas Low is Professor in Urban and Environmental Planning at the University of Melbourne, Australia. He is Founder and Associate Director of the Australasian Centre for the Governance and Management of Urban Transport (GAMUT). He is a member of the International Editorial Board of the journal *Urban Policy and Research.*

Fujio Mizuoka is Professor in Economic Geography at Hitotsubashi University, Tokyo, Japan. He is the chief of the Japan research team of GAMUT. He developed the conception of space subsumption, presented in the book he edited: *Geography of Economy and Society*, published in 2002 by Yuhikaku, Tokyo.

Patrick Moriarty is an Adjunct Senior Research Fellow in the Department of Design, Monash University, Australia. He is a partner in the Australasian Centre for the Governance and Management of Urban Transport (GAMUT).

Haixiao Pan is Professor and Director of Land Use-Transport Studies in the Department of Urban Planning, College of Architecture and Urban Planning, Tongji University, Shanghai. He has led many planning studies for Chinese cities. His research focus is the development of sustainable transport and transit oriented development, bicycle planning, and the impact of the Shanghai metro.

Julie Rudner is a Lecturer in the Community Planning and Development Program, La Trobe University, Bendigo. Through research, consulting and teaching, Julie explores children's views and experience of their environments, promotes cities as sites for experiential learning and supports children's citizenship through participation in planning and urban design. She seeks to understand how adults and children can expand their confidence and skills to encourage children's active exploration and engagement with their environments. She has more than 10 years' professional experience as a planner and consultant.

John Stone is a Research Fellow at the Swinburne Institute for Social Research and Lecturer in the Urban Planning Program in the Faculty of Architecture, Building and Planning at the University of Melbourne. His research explores the political, institutional and technical factors that support change to more sustainable urban transport.

Sophie Sturup completed her PhD *Managing Mentalities of Mega Projects* in 2011. She completed a Masters of Science in Geography at the University of Oxford and has a graduate certificate in public sector management from Flinders University. She is currently Principal of Sturup Consulting and undertakes research in GAMUT, and the Faculty of Architecture, Building, and Planning at the University of Melbourne.

Marcus Wigan is Professor Emeritus at Edinburgh Napier University, and a Visiting Professor at Imperial College London (www.mwigan.com). He has a broad background in transport, information systems and policy. As a GAMUT Partner from 2009, he has focussed on the need to alter the information power balances between community and government in planning and transport.

Carolyn Whitzman is Associate Professor in Urban Planning at the University of Melbourne. She is the lead editor of *Building Inclusive Cities: Women's Safety and the Right to the City* (Routledge, 2012) and author of *The Handbook of Community Safety, Gender, and Violence Prevention: Practical planning tools* (Earthscan, 2008). Her current research interests include: the policy implications of planning for healthy and equitable cities in a national and international context, planning interventions to increase independent mobility for children, and the development of participatory planning tools with women, children, and people with disability.

Foreword

This book is the synthesis of six years of research work at the Australasian Centre for Governance and Management of Urban Transport (GAMUT) at the University of Melbourne, Australia and its partner institutions. The GAMUT Centre was established following a successful competitive application to the Volvo Research and Educational Foundations by an international research group led by Nicholas Low at the Faculty of Architecture, Building and Planning at the University of Melbourne (http://www.abp.unimelb.edu.au/gamut/). Partner institutions included the Department of Urban and Regional Planning, Curtin University, the Urban Research Program, Griffith University, the Faculty of Economics, Hitotsubashi University, Tokyo, and the Department of Urban Planning, Tongji University, Shanghai.

GAMUT is unique in two ways. First, it is the only transport research centre in Australia explicitly linking transport with the environment and pressing for the change necessary to combat global warming and the diminishing supply of cheap oil. Second, GAMUT is the only research centre in Australia, and perhaps worldwide, specializing in governance and management of urban transport systems. Success or failure of transport systems depends significantly on how they are governed and managed, and also how they develop and adapt to new external challenges. Most importantly also, following the VREF model, the GAMUT Centre of Excellence (CoE), together with its sister CoEs around the world, has enabled a multidisciplinary and multi-perspective approach to future urban transport, an approach which is a particular strength of the current volume.

The Centre is one of a collaborative network of eight – soon to be nine, as of 2012 – Centres of Excellence in Future Urban Transport established by VREF. Others are located in the USA (two), UK, South Africa, China, India and Chile. The directors of the Centres, PhD students and staff meet regularly to discuss common issues every two years in Sweden and annually at other Centres. The VREF network provides an important opportunity for international learning and discussion.

The focus of the GAMUT Centre was at first specifically *urban* transport, following the VREF programme in Future Urban Transport. GAMUT's vision for the future is a transport system for cities that maintains the mobility that people need within the low carbon limits necessary to stabilize the climate system. Our focus is not only defining what that means – in adapted technologies – but also how to get there – how our public institutions and community perceptions must change.

GAMUT has made a significant impact in the field of future urban transport in three ways. First, GAMUT has provided a practical focus and unifying vision for research across urban planning, engineering, political science, public sector management and social geography. Second, GAMUT has introduced to debates on urban transport in Australia and the East Asian region a long-term global outlook and an independent platform for discussion distanced from the cut and thrust of daily media and political debate. Third, GAMUT has analysed significant operational means to achieve the vision of sustainable urban transport.

The problem we grappled with in the first five years of GAMUT's existence was governance and management: how to make the technologies of existing transport systems work better. Our focus was better machinery of government not better machines. We wanted to find ways forward to sustainable transport, and that project made us aware that solutions had to take account of the global context as well as local concerns. We identified a fundamental dilemma: on the one hand the health of the world economy and many societies depends on mobility of people and goods now extending worldwide. On the other hand this mobility is fuelled overwhelmingly by non-renewable carbon fuel whose emissions are damaging the planetary ecosystem. Since the dilemma of future transport is global in scope, a solution that fits one technology or one kind of city – the rich kind for example – but cannot be applied across the world, albeit in different forms, is not an adequate solution. Our research programme has gradually widened to include freight transport and sustainable logistics, and it became apparent that intercity and international movements could not be readily separated from the urban patterns of movement. This book is the latest (2012) of many publications by GAMUT partners. We hope to continue and extend our research programme into the future.

Nicholas Low
21 April 2012
GAMUT Centre for Sustainable Urban Transport

GAMUT CENTRE FOR SUSTAINABLE TRANSPORT
Governance and Management of Urban Transport
A Volvo Research and Educational Foundations Global Centre of Excellence
At the University of Melbourne

Acknowledgements

This book and the research it reports would not have been possible without the generous support of the Volvo Research and Educational Foundations, and its remarkable programme in Future Urban Transport of which GAMUT is proud to be a member (www.vref.se/). Through GAMUT, VREF supported the PhD research of Crystal Legacy, Sophie Sturup, Julie Rudner, Tim Petersen and Kate Alder. The Foundations not only provided generous funding for the GAMUT Centre but also mounted regular forums for interaction among all the VREF Centres of Excellence. By supporting Centres of Excellence VREF has created unique platforms, and an academic network, for interdisciplinary research on real world transport issues. The senior management of VREF always provided encouragement and useful advice during their visits to Melbourne. In this respect we would particularly like to thank the former Chair of the Board of VREF (and former Executive Vice President of AB Volvo), Arne Wittlöv, former Chair of the VREF Scientific Council, Bengt Kasemo (Emeritus Professor of Physics, Chalmers University, Sweden) and former VREF Board member Måns Lönnroth (former Managing Director of MISTRA, Sweden's Foundation for Strategic Environmental Research). VREF staff members Fabienne Niklasson, Peter Thormählen and Henrik Nolmark were very helpful to GAMUT in managing relations between VREF and GAMUT. We also warmly acknowledge the support of the OMEGA Centre for Mega Projects in Transport and Development at the Bartlett School of Planning, University College London, and in particular Harry Dimitriou with whom we worked closely to provide Australian case studies of mega urban transport projects. OMEGA provided financial support for the case studies and for the PhD research of Sophie Sturup. Carey Curtis provided support for Sophie Sturup at Curtin University in connection with the OMEGA project, and with Sophie Sturup and Nicholas Low wrote working papers for the OMEGA project. Imran Muhammad provided research assistance in the early stages of the OMEGA project.

Closer to home, the Australian Research Council provided funding for the research by Nicholas Low and Carey Curtis on path dependence in Australian transport policy, and – together with the Swinburne Institute for Social Research, Swinburne University, Melbourne – supported research by John Stone reported in Chapter 8 of this book. Terry Burke is acknowledged for his contribution to the research in Chapter 8. The New Zealand Transport Agency provided funding for the research on a benchmarking tool for planning of new public transport service patterns described in Chapter 9.

GAMUT itself depended on the collaboration of many scholars. Bill Russell was a Co-Director of GAMUT with Nicholas Low from its foundation in 2006 to 2011. Bill is a fountainhead of ideas and worked with Nick in 2005 to formulate the proposal for the GAMUT Centre and bid to VREF. He continues to inform and inspire our work. Bill has headed two government departments of the State of Victoria, Australia, and several research units. He was consulting editor for the *International Encyclopaedia of Public Policy and Administration.* He is co-author, with J. Shafritz, of the key international text on public sector management *Introducing Public Administration*, now in its 4th edition, published by Addison Wesley Longman (New York). Partners of GAMUT included Carey Curtis (Curtin University, Australia), Brendan Gleeson, Jago Dodson and Matthew Burke (Griffith University, Australia), Patrick Moriarty (Monash University, Australia), Fujio Mizuoka (Hitotsubashi University, Japan), Haixiao Pan (Tongji University, China), and at the University of Melbourne: Carolyn Whitzman, Sun Sheng Han, Kevin O'Connor, Marcus Wigan and Leigh Glover. Paul Mees was formerly a GAMUT partner and is now a senior researcher at RMIT University. He is internationally recognized for his work on public transport for dispersed cities, in for example the European Union's 2005 HiTrans project on improving public transport in medium-sized cities and towns (www.hitrans.org). Paul Mees's latest book *Transport for Suburbia: Beyond the Automobile Age* was published by Earthscan in 2010.

In this book Carey Curtis reports her research both for GAMUT and the Australian Research Council. The Curtin team included Jan Scheurer, Rachel Armstrong and Courtney Babb. Rachel Astle and John Odgers contributed substantially to research on path dependence reported in Chapter 6. Anthony Kent provided assistance in undertaking research for Chapter 7 – led by Leigh Glover. Chapter 10 by Carolyn Whitzman reported four years of research and action, with energy and ideas contributed by a large number of research assistants and research students, notably researchers, Lucy Pike, Jana Perkovic, Vivian Romero, Megan Worthington, and Dana Mizrachi, and PhD students Patrick Love, Julie Rudner, and Andrea Cook.

Importantly, Leanne Murphy and Danielle Hearn managed the day-to-day business of the GAMUT Centre, and liaison among partners and with the Faculty of Architecture, Building and Planning. Leanne and Danielle brought great charm and courtesy, as well as efficiency, to the demanding task of running the GAMUT office. We also thank their predecessors as managers: Kristen Sharp and Angela Munro.

Izumi Takeda (at Hokkaido University of Education at Iwamizawa, and founder of the National Railway Users' Union in Japan) was an important contributor to the Japan research. We also wish to thank and acknowledge the valuable work of the following, who assisted with the initial GAMUT bid to VREF: Mark Considine (then Director of the Centre for Public Policy at the University of Melbourne) and the late Jim Stevenson (then with the Department of Infrastructure, Government of Victoria).

The GAMUT Advisory Board has met regularly to offer valuable advice on transport issues and research, as well as giving important professional and intellectual support for the GAMUT Centre and its programmes. GAMUT gratefully acknowledges the contribution of the present members of the Board: Michael Kennedy (Chair), Glenyys Romanes, Janet Rice, Wendy Morris, John Hearsch, Michel Masson and Robert Fremantle. Also we wish to thank Dennis Cliche – former Chair of the Board – and former members, Max Lay and Peter Christoff.

The GAMUT programme is supported by honorary fellows of GAMUT: Paul Barter (Lee Kuan Yew School of Public Policy, National University of Singapore), Damon Honnery, (Associate Professor in Mechanical and Aerospace Engineering at Monash University), Susu Nousala, (Research Fellow at RMIT University), Glen Searle (Editor in Chief, *Urban Policy and Research*), and Jan Scheurer (Research Associate at the RMIT-NATSEM Centre, Australian Housing and Urban Research Institute, and at the School of the Built Environment and the Sustainability Policy Institute, Curtin University). Ms Chandra Jayasuriya, cartographer at the University of Melbourne, redrew diagrams and maps for publication. Michael Fell, Kat Holloway, Ginny Gilmore and the Routledge publication team took this book efficiently and helpfully through from a proposal to publication.

On a personal note, Nick Low thanks his wife Liz Low – Australian ceramic artist – for her continuing support during the preparation of the book.

Acronyms and abbreviations

ABS	Australian Bureau of Statistics
AFL	Australian Football League
ASPO	Association for the Study of Peak Oil and Gas
BITRE	Bureau of Infrastructure, Transport and Regional Economics (Australia)
BSTM	Brisbane Strategic Transport Model (Multi-Modal) (Queensland, Australia)
CFPTVC	Committee on the Future of Personal Transport Vehicles in China (China)
CoE	Centre of Excellence
CO_2e	Carbon Dioxide Equivalent
CIM	Children's Independent Mobility
DCCEE	Department of Climate Change and Energy Efficiency (Australia)
DoI	Department of Infrastructure (Victoria, Australia)
DPUD	Department of Planning and Urban Development, Western Australia
ECMT	European Conference of Ministers of Transport
EIA	Energy Information Administration
EJ	Exajoule = 10^{18} joules of energy
EU	European Union
CBD	Central Business District
GAMUT	Australasian Centre for Governance and Management of Urban Transport
GDP	Gross Domestic Product
GHG	Greenhouse Gas
GIS	Geographic Information System
GVRD	Greater Vancouver Regional District (Canada)
GVTA	Greater Vancouver Transportation Authority
HFCV	Hydrogen Fuel Cell Vehicle
IC	Internal Combustion
IPCC	Intergovernmental Panel on Climate Change (UN)
IEA	International Energy Agency
JNR	Japan National Railways

JR	Japan Railways
LPG	Liquid Propane Gas
LRT	Light Rapid Transport (Singapore)
LTA	Land Transport Authority, Singapore
LUTI	Land Use-Transport Integration
MULUTT	Modular Urban Land Use and Transport Tool
MUTP	Mega Urban Transport Project
MRT	Mass Rapid Transit (Singapore)
$MtCO_2e$	Megatonnes, Carbon Dioxide Equivalent
NIDL	New International Division of Labour
OECD	Organisation for Economic Cooperation and Development
OPEC	Organization of the Petroleum Exporting Countries
PACTS	Parliamentary Advisory Council on Transport Safety (UK)
PPP	Public–Private Partnership
PTA	Public Transport Authority
PTC	Public Transport Council (Singapore)
RDA	Redevelopment Agency (or Authority)
RE	Renewable Energy
RTS	Rapid Transit System
SACTRA	Standing Advisory Committee on Trunk Road Assessment (UK)
SLC	Supplier Logistics Centre
SNAMUTS	Spatial Network Analysis for Multimodal Urban Transport Systems
SOE	State-Owned Enterprise
S$	Singapore dollars
TOD	Transit Oriented Development
TGV	Train Grande Vitesse (High Speed Train)
TMR	Transport and Main Roads (Queensland, Australia)
UNEP	United Nations Environment Program
USGS	United States Geological Survey
VKT	Vehicle Kilometres Travelled
WAPC	Western Australian Planning Commission
WBCSD	World Business Council on Sustainable Development
WCED	World Commission on Environment and Development
WTO	World Trade Organization

Part 1
The global dilemma

1 The dilemma of mobility

Nicholas Low and Kevin O'Connor

Introduction

Transforming Urban Transport confronts head on the dilemma faced by a world wedded to mobility: the danger of continuing along the fossil-fuelled transport path and the real difficulty of replacing private vehicle solutions in time to avoid dangerous climate change. There are major barriers to change and it takes time and strategy to overcome them. But there are also solutions and ways forward, some of which we explore in the chapters to follow. It is the purpose of this chapter to sketch out the argument of the book. Many of the ideas are not new, but the way we present the problem and bring together strategies to overcome it represents a new perspective and reveals more about the necessary social and institutional change than has hitherto been attempted.

In earlier work (Low and Gleeson, 2003: 17; Low, 2002) we have argued that transforming transport will occur through a process of 'ecosocialization' following Karl Polanyi's 'Great Transformation' in which the social contradictions of capitalism were partially resolved through a movement of social protection that culminated in welfare states (Polanyi, 1957). In the same way a movement of environmental protection is attempting to resolve the environmental contradictions of capitalism. There is evidence from around the world that this is happening. But will it move fast enough? In truth the task of transforming transport on a global scale is vast. We see the same cities cited again and again as evidence of change, and as models of good practice – for example, Portland, Vancouver, Curitiba, Zürich, Freiburg, Copenhagen, and sometimes Seoul and Singapore – yet none of the transport systems even of these cities can measure up to a rigorous analysis as 'sustainable'. And the vast bulk of cities and new urban development worldwide remain determined by the paradigm of automobility. Even the impoverished nation of Pakistan is building roads and flyovers to accommodate the tiny minority of aspiring motorists (Imran, 2010). The dilemma is so profound and the roots of unsustainable transport run so deep that both urgency and a degree of pessimism are warranted if we are to view the problem of change in its true light.

The rise and fall of automobility

We characterize the vision driving transport policy during most of the twentieth century – and still persisting – as *automobility*, a conception of transport that came to infuse thinking about cities and their future worldwide. In the latter part of the twentieth century, city building proceeded on the assumption that the spatial relationships among the various activities going on in urban regions were best left to the individual choices of persons and firms under the discipline of a land market. The need to move between activities, for so long organized by collective transport systems, could now be met by individual mobility.

The form of transport that most perfectly matched the individualistic ideal of the free market was the private motor vehicle. Private cars greatly increased the choice of where to live and work, and freed labour from the tyranny of proximity to workplaces. As a mode of travel the car combined speed, flexibility and control over personal space (Pooley et al., 2006). For firms, private trucks could deliver goods 'just in time' to meet demand wherever, whenever they were wanted, unlimited by timetabled services at railheads or ports. In this way the crude rigidities of collective transport could be avoided. In real terms too, this form of mobility became cheaper as the cost of cars and trucks, and the cost of fuel, fell steadily – despite oil price spikes in the 1970s. The consequent flexibility of movement provided a flexibility of locational choice for firms and individuals in tune with the flexibility of markets for goods and services.

A knowledge industry employing hundreds of thousands of consultants and scholars was built around making and justifying the infrastructure needed to allow the ideal of 'free movement' to be pursued. Engineers not only perfected efficient road design but also developed the traffic, land use and travel behaviour models that would come to define the problem to which engineering mega-projects were the solution. The 'project sublime', able to slice through mountains, lightly span great distances, keep traffic flowing and tunnel between the pipes and foundations under cities impressed politicians and the public. All transport problems, it seemed, would yield to engineering solutions – while defining the problems in the terms that engineering would solve.

But these compelling ideals turned out to be a mirage. Although it seemed that transport relied upon individual choice, cars and trucks run on publicly constructed roads. Pure market principles were violated in the spatial unevenness created by the selective network effects of the location of the highest quality roads and railways. Rights to travel entail 'specific often highly embedded and immobile infrastructures' (Sheller and Urry, 2006: 210). Collective transport continued in use, filling the gaps where individual transport failed. Traffic congestion increased. Urban environments filled with vehicle fumes and noise, and in some places vast areas of parking lots took over the public space. Land use planning controls limited the locations of some activities in most cities. Road traffic caused deaths and injuries worldwide on a scale greater even than the scourge of malaria. Finally, individual transport was only 'cheap' when its external impacts were ignored. So the market-technical ideal was never really approachable. Today, with the looming fact of global warming and the need to stop burning fossil fuel, and with the rapid depletion of oil – the source of cheap transport – the mirage has dissipated further so that individual motorized mobility will need to be re-thought as a means of connecting activities in urban regions in the twenty first century.

The tension within cities between the ideal and the reality is replicated at the global scale. The neo-liberal governance of the global economy requires that the location of productive activities be determined by flexible supply and demand forces in the global market. Mobility provides a way to take advantage of these forces. So production now involves global supply chains connecting the activity of large and small firms often over vast inter-continental distances. Cheap mobility has solved the problem of distance, not only between producers and consumers but also between different producers and among the production facilities within corporations. The health and efficiency of the global economy, it can reasonably be said, now hinges on the cheap mobility of goods. Here too global warming and oil depletion will force the global economy to face an uncertain future. And insofar as the global economy is an economy of cities, the problems of global mobility intersect with those of urban mobility.

The idea of sustainable transport

As the ideal of *automobility* entered its long decline, a new ideal model started to take shape characterized as *sustainable transport*. Instead of starting from the premise that perfect mobility was possible and desirable within safe speed limits made available by current vehicle technology, sustainable transport starts from the premise of reducing the social and environmental damage caused by motorized mobility. This understanding recognizes that oil production will peak and decline, costs of fuel will rise, and that there will be growing demand for better public transport of people and goods. However there remain serious conceptual problems, and problems of measurement of 'sustainability'. The new paradigm of sustainable transport and its governance – both global and local – are discussed in Chapter 4.

Dissatisfaction with automobility, or 'hypermobility' (Adams, 1999, 2000) stemmed from the actual loss of quality of urban environments and worries about urban sprawl, the destructive effects on rural environments caused by road building, the discovery and confirmation of 'induced traffic' (SACTRA, 1994, 1999; Goodwin, 1996), fears for the consequences of comprehensive plans for cities served by networks of motorways (Sachs, 1992; Davison with Yelland, 2004). In later years, to these fears were added the global problems of climate change and peak oil – the tendency of world oil production to reach a peak and then decline (Campbell, 2003; Betsil and Bulkeley, 2007; Chapman, 2007; Aleklett et al., 2010).

In an attempt to define alternatives to automobility much has been written about the idea and practices of sustainable transport (Whitelegg, 1997; Newman and Kenworthy, 1999; Low and Gleeson, 2003; Schäfer et al., 2009; Schiller, Bruun and Kenworthy, 2010). The OECD and even the car companies themselves have adopted the concept of sustainable travel or sustainable mobility (OECD, 1995; WBCSD, 2004). Plainly 'sustainability' embraces a multitude of conflicting interests. However, there are three aspects of the problematic on which all analysts seem to agree: the global scope of the problem, the deeply entrenched character of automobility and the need to change. It is the question of institutional change in varied contexts but of global scope that is the main theme of this book. Given the acknowledged urgency of the need for the transport sector to change, how is this to be achieved? Indeed, can it be achieved in time to avoid the most dangerous consequences of global warming?

The book addresses this question via an analysis of three themes: the global dilemma of mobility, the institutional dimensions of change, and local strategies to induce change.

The dilemma of mobility (Chapters 2 and 3)

This theme poses the dilemma shaped by practices of mobility deeply entrenched in societies and cultures worldwide, now juxtaposed with an acknowledgement of the need for massive and rapid cuts to emissions from the fuels that have driven this mobility. Understanding the dilemma, we argue, is the first step towards changing the assumptions and developing strategies.

The institutional dimensions of change (Chapters 4 to 6)

The dilemma outlined above is expressed in the institutional and cultural structures that have facilitated and enhanced the current mobilities. Hence change will only occur once the assumptions embedded in these structures are identified and understood, and the structures themselves change. Their political, social and economic persistence – the common norm – acts as a barrier to change in most societies.

Strategies for local change (Chapters 7 to 11)

Strategies for change will be multiple and variable. Strategies must start from today's realities. To illustrate that, a selection of practical strategies is discussed. These include the governance of transport planning in dispersed and concentrated cities, the analytical techniques aimed at indicating transport need, the policies to induce change in mobility for the child, the political mobilization needed to promote change, and the way new knowledge may be mutually discussed and shared.

Shifting the world to put sustainable transport into practice and replace the dominant global model of automobility requires us to go beyond stating and restating the principles of sustainable transport and giving good examples of practice – and in a global context they are in reality all too few. If we are to change our habits of mobility in the interests of stabilizing the global climate and coping with reduced oil supply we should be clear about the basis of those habits, and find ways to help them change.

The significance of mobility

Geographers and economists have long recognized the 'collapse of spatial barriers' over time and the important role of the private vehicle – passenger and goods – in shaping the development of cities in the twentieth century. The significance of that role was captured by Clark (1996: 100) when he observed that, 'the ability to participate in an urban way of life is largely independent of location and is open to all'. There is nothing new about this outcome. Personal mobility has been growing steadily in the developed world for centuries, and that is being reprised in the emerging economies in step with the development of technology and the broadly based economic development associated with industrialization. Pointing out the institutional as well as material determinants of mobility, Gogia (2006: 359) observes, 'the conditions of globalisation have facilitated the movement of a record number of people; individuals who are crossing international borders for work, leisure, safety and security'. Within cities, just as urban collective transport provided mobility for a wider and wider set of social groups, so too the mass production of the private car offered a type of personal mobility to the whole adult population which was formerly the privilege of the rich. Bringing mobility back to the core analysis of social life means acknowledging that mobility has at least as important a role to play in shaping social life as 'place', location and political boundaries, overturning what Cresswell and others term the 'sedentarist' basis of geography and politics (Cresswell, 2006; Kaufmann, 2005).

It is under this sedentarist assumption that mobility is conventionally seen as an epiphenomenon, a 'derived demand', a mere concomitant of something else such as urbanization or globalization, or a means to other activities conducted in place. Banister (2005: 224) points to the tension between transport seen as a negatively valued activity – a derived demand – and an activity with intrinsically positive values. To understand the mobility dilemma we need to bring physical mobility into the foreground. What does being able to move around quickly and freely do for the way society works? The shadow question is of course 'how will society cope if its members cannot move around in the same way?' We need to understand what the raw fact of vastly increased physical mobility – in simple terms the ability to travel more often and further afield – what Whitelegg (2003: 118) terms 'the distance-intensive lifestyle' – means for society. We need to know more about this in order to begin to comprehend what a change in mobility might mean.

To understand the effect of mobility on society we need to recognize some of its key features. Our research leads us to define mobility in terms of three aspects. These are measures of *distances travelled*, *frequency of travel* and the *proportion of the population who are travelling* summed up in the diagram in Figure 1.1.

The essence of modern mobility is that high proportions of the population are travelling frequently, often over long distances. That is in contrast to the long-term historical record in which very small proportions of the population made infrequent long journeys, while the majority of the population travelled short distances frequently. Thus in the world today the experience of mobility has shifted, and is still shifting, towards the upper right hand side of the diagram. To understand mobility we need to understand this new pattern. There are a number of possible explanations that we will discuss under the conventional rubric of production and consumption.

Industrial production

Economic geographers once emphasized capital mobility as a key element of globalization. More fundamentally, it is the mobility of goods in global supply chains that is the working face of globalization. Mobility is so fundamental to production that we can perhaps think of the 'embodied mobility' in almost everything that is produced. Thomas Friedman (2005: 415–17) provides an example by outlining the many inter-continental links that were needed to produce and deliver his Dell notebook computer (Box 1.1).

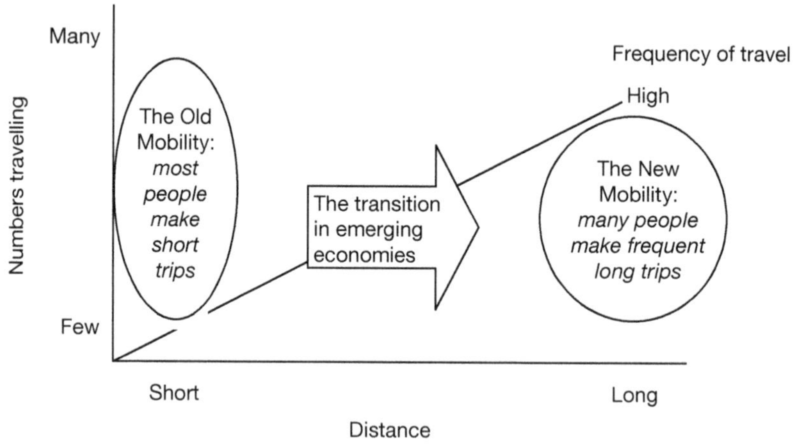

Figure 1.1 Distances travelled, frequency of travel and the proportion of the population who are travelling

Box 1.1 Global trade and the Dell notebook

Thomas Friedman (2005: 415–17) tells the story of the many national and intercontinental links required to bring his Dell electronic notebook to his desk. Dell has six factories around the world spread over four continents. Surrounding every factory are supplier logistics centres that have stocks of parts supplied from all over the world. Key elements of the notebook's hardware come from the Philippines, Malaysia, Taiwan, Germany, Mexico, Singapore, Thailand, India, Israel, and of course many different cities in China. Friedman lists them in great detail. Dell notebooks are constantly being updated and are completely redesigned about every twelve months with new hardware and software components. The key elements of the notebook come from all over the globe. This way if one supplier fails to deliver, the part can be sourced from somewhere else: 'Every two hours, the Dell factory in Penang sends an email to the various SLCs [supplier logistics centres] nearby, telling each one what parts and what quantities of those parts it wants delivered within the next ninety minutes – and not one minute later. Within ninety minutes, trucks from the various SLCs around Penang pull up to the Dell manufacturing plant and unload the parts needed for all those notebooks ordered in the last two hours' (ibid. 415).

Here is an example of the detailed operational planning of mobility. Friedman's example illustrates the logistical logic in which labour skills in many places around the world are coordinated in production. The Dell notebook is just one of the thousands of commodities produced that flow through centrally planned global supply chains of this sort.

Global supply chains reflect the organization of production and, in particular, the way in which efficient supply allows the best use of specialized knowledge. This process begins with design and finance, is carried through with actual production, and ends with marketing and distribution. Mobility has allowed the concentration of each of those activities at its 'best' place – in terms of least cost, least risk and greatest opportunity – on the globe. So the thinking about design and finance takes place in and around the global cities of economically advanced countries, with the skills of managers and related professionals moved when needed by air travel to production locations. Production sites are scattered to places with particular skills in particular components – as the Friedman example of the Dell notebook illustrates. Final distribution and consumption are located in other countries and continents. Marketing experts travel by air to discuss approaches in particular markets, and goods flow to retailers via sea, air, road and rail, via gateway sea and air ports, their movement managed by skilled logistics companies.

Hence mobility allows for specialization in production. In turn it can be linked to creativity and innovation. That process draws not only on local skill and knowledge but also contacts and networks that connect people.

The latter have been analysed in a construct called 'creative cities' (Florida, 2005; Cooke and Lazzeretti, 2008) or 'knowledge cities' (Carrillo, 2006), where the local circumstances seem to be a critical element in what makes for innovation and an expanded view of the world. However, creative cities often have busy airports, as contacts and networks extend beyond the local to draw in the new and the different, and connect the creativity of design with the actual reality of production. So, mobility is associated with the use of knowledge in creativity and innovation, now seen as the critical dimension in modern economic development.

Most production, however, does not involve these global ties. Much more is local and regional, dealing with the consumption needs of local populations. Here, too, mobility has a critical role to play in supporting technical specialization as activities like electrical work, plumbing and building move around large markets with people carrying out their specialized skills. To them can be added many other professionals who cannot rely upon consumers to come to them, but rather move around a market, for example medical services, where specialists spend short periods in several different locations.

Personal consumption

Just as production is a powerful driver of mobility, so too is consumption. Let's anchor the idea of frequency and journey length in everyday reality by way of an easily recognized example:[1]

> Emma and Max are a moderately wealthy Australian middle-class couple. Up until a few years ago they lived in an environmentally attractive suburb on the outer fringe of a large city. Max worked in the CBD, Emma worked some ten kilometres away in an adjoining suburb. Both used separate cars to get to work – and, as soon as they could drive, their two children also bought cars. Now that their children live separate lives, Emma and Max have moved to an inner city tenth floor flat with a fine view. Max cycles to work, where he and his colleagues have notched up some twenty intercontinental air journeys associated with their business in the past four years. In spite of their inner city location, Emma's and Max's single fuel-efficient car – made in South Africa – clocks up 15,000 kilometres per year, because Emma and Max have a holiday house near a beautiful stretch of beach about two hours drive away from the city. Max, who grew up in England, still has many friends there, and they fly to Europe nearly every year. They go walking in the French Alps to experience a different culture, landscape, flora and fauna. Part of the experience is sampling cheese made from the milk of cows that graze on flowery alpine meadows. In 2009, however, they flew to Darwin to join a cruise to Broome around the Kimberley coast. Down at the beach house they like to be reminded of French culture by eating Beaufort cheese made in the

> alpine Tarentaise valleys and drinking white wine with a few drops of sirop de pamplemoussse. The latter is made in the village of Nuits St George in Burgundy from pink grapefruit imported from Morocco. These goods enter Australia in the multi-coloured container ships that berth briefly in Melbourne on their global journeys. From the docks the goods move by truck to warehouses, supermarkets and shops. Most of the clothes on their backs and much of their flat pack furniture at the beach house are made in China with energy supplied by power generated by burning Australian coal shipped north by sea in a steady line of diesel powered bulk carriers.

This vignette, with infinite variations of location and diversity of destinations, is increasingly typical of urban life in the developed world. Cities of the emerging economies will not be immune to such mobilities, even though at present large parts of their societies are shaped by very different physical conditions, modes of travel and proximity relationships (see Brugmann, 2009). The mobilities enjoyed by Emma and Max reflect the discretionary lifestyle choices they have made – easy for puritanical environmentalists to scoff at, but these choices need to be understood so as to bring mobility to the centre of analysis of future urban transport.

These choices reflect an interest in diversity and variety, and a value placed on 'out there' or 'the other' above the local and known. That value can be expressed in types of food as in the case above, but also in choice of education, where many students now travel long distances within countries and between continents, even though local supply is available – a fact that provides Australia with a significant export sector: tertiary education. More fundamentally, as in the case of furniture and clothing in the example above, consumption behaviour reflects the fact that locally made goods are no longer available as global trade has driven local manufacturers out of business, because there are better, cheaper goods made elsewhere.

Seeing mobility as a consumption item recognizes that it equips lives with both essentials and luxuries, allows a wider choice of residence and employment, opens up the range of recreational possibilities, and provides access to friends and social activities throughout the city and the world. Looked at in this way mobility is a potent force of human desire.

The challenge of discovery

The intensity with which mobility is embedded in our social and economic life raises the question of whether there is an even more deep-seated way in which mobility is a human desire. Is it by mere accident or necessity – for example, food shortage – that the human species spread all over the globe very early in its evolutionary history? Or is there an innate human tendency to explore, to seek out the new, to value variety and diversity? In historical times has exploration been driven purely by material considerations, for instance to find new products to trade, or was exploration driven by the very challenge of discovery for its own sake? Legend has it that the sailors setting off with Vasco da Gama to find a sea passage to India believed that the sea boiled just off Cape St Vincent in Portugal. Yet they went. Now teenagers set off to sail round the world single handed. It could hardly be said that putting a man on the Moon was driven by economic necessity. Now we have space travel being offered to tourists. Far more significant for mobility, the challenge of discovery, once the property of the very few, is now available to many. Discovery has been democratized. Once, geographical discovery was the work of a few people travelling infrequently. Today very large numbers of 'adventure tourists' are drawn much more frequently to the new and the different. Yet, paradoxically, children of competent age are being prevented from discovering their own neighbourhoods by themselves, as we point out in Chapter 10.

Of course we need to acknowledge the strength of home and place, and perversely, in some cases, the desire to protect territory and keep others out – like the gated communities springing up in suburbs, and local zoning controls. There are certainly those who wish to make the whole of Australia a gated community and stop the unregulated migration of asylum seekers – in Australia: 'boat people' – or in the USA to keep the Mexicans out or send them back. There is a political language that designates the suppliers of travel to refugees as 'people smugglers' and not travel agents, and refugees as 'illegal migrants'. But then the cultural distinction of place and territory is also something that makes 'other' places attractive and different. Europeans whose governments have zealously protected cultural values connected with place – as in the rules of food production related to 'terroir' and 'appellation' – worry that their lands may become theme parks whose primary function is to serve the tourist explorers from elsewhere. In recognizing this we are just adding another layer of complexity to the role of mobility in our world.

These three linked aspects, consumption, production and the challenge of discovery, provide a way to understand the high-frequency, long-distance travel of a growing proportion of the population in many cities and countries, and over time the steady move along these axes by populations in emerging economies. Hence mobility is not a derived outcome measured by transport statistics. It is a fundamental component of a way of life, one that is embedded in most of our social, economic and cultural endeavours. To say all this seems to be stating the obvious. Yet this 'obvious' is actually not much stated, and therefore needs to become the primary focus of scholarly attention.

What is perhaps most interesting is not so much the raw fact of mobility as its connection with the range of other social and technical phenomena outlined above: with telecommunications which facilitate the organization of mobility, with social interaction, with the activities of production and consumption, culture and symbolism, with social class, the regulation of capital and labour markets, the economic prospects of populations far distant from one another, and the future form of cities. The physical mobility of goods and the handling of 'freight' in cities will affect their future form, in many cases pushing industrial and commercial land uses in an opposite direction from that more compact form which is expected to reduce the need to travel. The place of mobility as the foundation of our transport systems, with its impacts on urban development, is discussed in Chapter 2.

The mobility dilemma is formed from the way we have organized our mobility over the last fifty years, and the way this form of mobility creates serious – indeed life threatening – problems and thus has a short future. We turn now to the second 'horn' of the dilemma, the real damage caused by the widespread pursuit of mobility.

The unaccounted costs of mobility

There have been enormous costs to the worldwide growth of mobility. Some of these costs are discussed in Chapter 3. The costs are social and environmental, and many of the individual costs are very unevenly distributed. For instance, in the places to which individual motorized mobility is now spreading, it is the non-motorized majority that bears most of the 'accident' trauma, while the motorized remain a small, if growing, minority (Ameratunga et al., 2006). Oil extraction has destroyed the environments and social structures of people in countries like Nigeria (UNEP, 2011). And, of course, the impact of climate change, felt as unpredictable local weather events, displaces the cost of mobility on to those who are least mobile and who cannot move away from threatened places, on to other species and on to future human generations.

The urban mobility of private vehicle transport today consumes more than 1.3 million lives per year and disables many more. Toxic pollution from vehicle exhausts almost certainly kills large numbers prematurely from cancers and bronchial diseases – vastly more than are damaged by radiation from nuclear accidents or terrorist attacks. Yet this social carnage has been borne as a cost of 'development' and become normalized. Car crashes are framed in terms of individual responsibility and individual tragedy, not social dysfunction. Morbidity from mobility is masked by the general improvement in health that comes with prosperity. Traffic congestion in cities can be managed in various ways, though it remains unsolved. But today mobility faces a global threat that cannot be absorbed or managed by society. The threat comes from the exhaustion of the potential of fossil fuel that has made today's mobility possible.

We tend to think of climate change and the peak of oil as two separate problems emanating from the present form of mobility. But they are deeply interwoven in complex ways, and the resultant congeries poses a threat to global society and economy. The mobilities discussed above depend on the cheap and convenient availability of fossil fuel for transport. It is the cheapness and availability of fossil fuel that is now under threat. The threat comes from two directions. First it is now well understood that the climatic cost of burning fossil fuel will have to be paid. If there is no change in mobility it will be paid in some combination of destruction of the biosphere and the monetary price of adaptation. The former has an economic cost that is unquantifiable, perhaps not in principle but because we do not yet know exactly what will be lost from biosphere destruction – perhaps our global civilization – and we probably will not know until it has happened. We do know, however, that the cost will be very unevenly distributed geographically and socially, a concomitant of which is likely to be loss of global security, which will in turn impinge on mobility. The price is very time dependent. If we pay more sooner, we will pay less in total (Stern, 2006; Garnaut, 2008). But this fact poses an acute difficulty for political systems structured to respond to short-term issues. It is also challenged by the span of human life. Discounting the future is not just an economist's fable. We all do it as a matter of common sense. But it is a habit we may have to give up if there is to be a future for the biosphere.

The second challenge is from the increasing scarcity of the most convenient fuel: oil. The theory of peak oil is rapidly being proved true by experience. Production from currently exploited oil fields is peaking, and no large-scale replacement fields are being discovered. Demand driven by the insatiable desire for mobility is growing. Unsurprisingly the price of oil is growing commensurately, temporarily stemmed only by the expectation of reduced economic growth resulting from the global financial crisis. There are unconventional sources of oil. But the most available is also the dirtiest: the Athabasca tar sands in Alberta, Canada, whose exploitation is on track to make Canada replace Australia as the world's climate change arch-villain. Deep sea and polar oil probably exist but both are expensive and environmentally damaging to extract. The environmental disaster of the explosion of BP's Deepwater Horizon well in the Gulf of Mexico illustrates the magnitude of environmental risk, and this was not even located in a particularly challenging environment. There is also petroleum gas in various forms whose supply will also rapidly peak if it comes to substitute for oil, and there is coal which is in no danger of running out but is more damaging in combustion than oil, and expensive to adapt for the internal combustion engine.

This brings us to the question to which Chapter 3 is addressed: can the dilemma of mobility be resolved by changes to vehicle and fuel technology? To answer this question the research examined what would be a necessary global response in the transport field to reduce emissions of greenhouse gases to levels estimated to produce no more than 2 degrees Celsius of warming above pre-industrial levels. The clear answer is that known and available technologies cannot accomplish this. Therefore we need to re-think many of our consumption habits and production techniques.

This then is the dilemma: continue and create chaos, conflict and environmental degradation, or change and live in a different world. But can we change? This book signals the need for a shift in direction of the social technology of mobility; from a form dominated by automobility to a form in which collective and active – essentially using human energy – transport become the dominant modes of transport in cities.

In Chapter 4 the book moves from the dilemma of mobility to explain the need for a new set of basic principles for sustainable urban transport. In the first part of the chapter, the iterative processes by which automobility developed and took hold in the rich industrial nations is explored: a circulating process ranging through public aspirations, interest formation in consumption and production, infrastructure building, construction of expert knowledge together with a corps of experts, solution and problem definition, technical opportunities and government responses. The origins of the idea of sustainability are then discussed, and the need to restore meaning to the term 'sustainable'. Principles of sustainable transport are set out which express the core idea of sustainability: broadly stated – living within environmental limits. Then the practices are described which have been proposed as most likely to implement the principles.

Finally, in the light of the positive qualities presumed to flow from automobility, the positive quality is considered that flows from 'sustainable transport'. This is 'quality of place': safe and attractive streets and public spaces. Automobility destroyed quality of place. Sustainable transport offers to restore it. It is argued that this is not the same as 'sustainability' but rather a co-benefit that is necessary in order to engage the public in the struggle for sustainable transport. The idea of safety, and especially child safety enabling independent mobility, is examined again in Chapter 10.

Under the assumptions of automobility, transport planning remains fixated on positivist analytical frameworks to assist decision makers to 'solve' transport problems, while at the same time defining what those problems are (Banister, 2002: 22). Vigar's work shows how technical assessments have often been used to frame transport debates within 'schemes' and 'hard' engineering projects with their focus on road corridor and network capacity, congestion and travel speeds: under the assumptions of automobility (Vigar, 2002, 2006). If there is to be a change of assumptions towards sustainable transport, it is necessary to open up the scope of transport assessment to include questions of equity, access, oil vulnerability, climate change and other emergent concerns. Foreshadowed here is the need for the new kinds of decision tool explained in Chapter 9.

The new set of assumptions emphasizes collective transport, building on known methods, and shifting investment to provide optimum conditions for safe active – non-motorized – travel over short distances for those able to do so. For reasons discussed in Chapter 3, technological optimism is problematic. And effective governance, linking global with national action, is a crucial ingredient of the required transformation.

Institutional dimensions of change

Even if it were technically possible to deliver low carbon mobility in and between cities, change of the form of transport embedded in global and national systems of production, consumption, exchange and distribution does not yield simply to rational analysis and clear statements of alternatives. Barriers exist in the institutional behaviour regulating production, and in personal and social habits of consumption. The existing physical infrastructure for mobility itself imposes a barrier to new systems by structuring cities in ways that confirm existing modes of mobility (Woodlief, 1998).

The physical structure of cities changes only very slowly. But that structure is also being continuously created at the margin by the action of governments and markets – much as a snail generates its shell as it grows. Transport infrastructure in a very real sense shapes the city. How people in government think determines the transport response to mobility demand. In Chapter 5, we examine change processes under the constraints on investment in transport systems at global level. The world economy in different periods is subject to different consensual sets of assumptions about how capital should be regulated. This was first postulated by European theorists (for example, Aglietta, 1979; Lipietz, 1984; see also Jessop, 1990) who drew attention to the shift from the 'Fordist' regime of strong government intervention in the economy – though in varying degrees in different countries – to the 'post-Fordist', now recognized as the neo-liberal regime of the 'Washington consensus' with its emphasis on deregulation, liberalization and the virtue of market-based governance.

In Chapter 5, we go beyond simply describing the regulatory strait jacket of governance under neo-liberalism. It is argued that breaking out of the mode of regulation is driven by community and class struggles. Two case studies in Japan show how neo-liberal orthodoxy was to some degree deflected by community action. In Chapter 4, we pointed to the need to change the set of assumptions underlying urban transport. But change to such major investments as transport infrastructure only occurs through response to real social and political pressures, and the struggles in which various socio-economic classes and groups are involved. Therefore, in order to achieve efficient, user-friendly and sustainable systems of transportation in societies – the model of *sustainable transport* – it is necessary to understand the politico-economic processes that influence shifts in transport policy (see Dudley and Richardson, 2000 for a detailed analysis of change in UK transport policy). Research reported in Chapter 5 attempts to identify the 'critical junctures' or real turning points when the dominant political body comes to the decision to shift investment patterns under the pressure of social crisis. Case studies from Japan are used to substantiate the argument.

Neo-liberalism is a widespread belief system, but it is insufficient to explain the regulatory behaviour of many governments, for instance China, Singapore and Malaysia. These countries appear to be following a Fordist or neo-Fordist institutional path, with strong government intervention and planning of both road and public transport infrastructure. And it cannot explain the considerable differences in detail among countries operating under the rubric of neo-liberalism. So in Chapter 6 we turn from the global to the national and sub-national levels of governance, and analyse the institutional barriers that appear at these levels.

We examine three frameworks for understanding the phenomenon of *policy stasis* – the inability to change the policy position in the face of new external circumstances and existential threats. The research discussed here draws on empirical investigations and case studies. The first framework is that of 'path dependence'. The theory of path dependence has been used to explain the persistence of policy, and resistance to change, over a wide range of policy domains and geographical areas (see Pierson, 2004; Low, Gleeson and Rush, 2005; Low and Astle, 2009; Curtis and Low, 2012). Instead of responding to consumer or citizen demand, it is argued, the main features of products, means of production or current policies are determined by certain critical junctures – as in regulation theory – in the history of their development, and reinforced in a variety of ways by supporting organizations engaging in apparently rational arguments on public record. Path dependence theory requires an examination of two related sorts of path. The first is the path set down by the weight of physical infrastructure: how the form an urban transport system takes over a period of time shapes the future of a city. The second is the institutional determination of policy. The institutional field can be further divided into the organizational power of the agencies making transport policy, and the discourse and language that define the transport problem for policy makers – how policy is shaped and supported by arguments, perspectives and storylines.

The second framework is that of 'art of government' based on Foucault's theory of governmentality. In this ontological framework the 'being' of those engaged in making transport mega projects is generated by an amalgam of technology, rationality and knowledge used to produce projects. In other words, people act in accordance with how the world occurs to them, how they see their place in the working world, which in turn encompasses the possibilities of behaviour. Being in a project imposes bounds upon behaviour that does not fit comfortably with the solutions to certain 'problems' that occur during the development and construction of projects. It is the existential boundedness of behaviour within the world created by mega projects that leads to the problems observed by Flyvbjerg et al. (2003), for instance cost and time overruns, as well as the inability to identify problems beyond that to which the megaproject provides a solution. In a neo-liberal world in which government planning has all but disappeared, the substitution of project bundles for plans has made issues of wider scope very difficult for government to address in a problem-solving manner.

The third framework concerns the knowledge of risk. Children in Australia today grow up in the belief that they are safe travelling in cars. When children start to become capable of independent mobility around the age of eight they are, in fact, restricted within a cocoon of 'safety' by their parents. Instead of, for example, using public transport, bikes or walking to school, they are driven to school by car. They are driven to sporting and recreational activities by parents. Over the last thirty years children's independent mobility in Australia has become increasingly diminished, as it has in Britain and the USA. There are pressures on parents to keep children safe from 'stranger danger' and road traffic. The media report frightening incidents of attacks on children and road traffic accidents involving children and pedestrians. Children appear to their parents to be highly vulnerable, fragile creatures, and parents imagine what might happen to them and what bad parents they will appear if bad things happen to their children. Much depends on how risk is construed in parents' minds. Yet 'feeling safe in cars', and being protected from experiencing the real risks of mobility as children, may have unintended consequences. When children become teenagers and learn to drive, and when their inhibitions are released by alcohol, they can experience the thrills of speed without feeling the dangers. Thus in Melbourne, on 18 January 2010, a 19-year-old boy and five of his mates who had been drinking at a party roared off in a powerful car, later lost control of the car at high speed and crashed into a mature oak tree by the side of the road, uprooting the tree, demolishing the car and instantly killing five of the car's occupants (*The Age*, 18 Jan. 2010 p.1).

In one sense risk is the fear that negative consequences may follow a certain course of action or policy. The opposite pole of risk, in this sense, is opportunity. Here, however, Rudner offers a more nuanced threefold conception of risk as a relationship between an individual's intent – what a person wants to do, what the environment affords in both opportunities and dangers, and the person's capacity to use the environment in the intended manner. This relationship is particularly illustrated in relation to children's independent mobility, to be further discussed in Chapter 10. Policy professionals and the lay public acting together can generate an understanding of risk focused on negative consequences that promotes policy stasis rather than experimentation, opportunity and capacity building, and thus feeds path dependence.

In concluding Chapter 6, two investigations of metropolitan planning are discussed. The first examined the barriers to implementation that result from lack of ownership of the plan by key players. 'Enlarged thought' (Arendt, 1978), necessary to grapple effectively with the mobility dilemma, can be assisted by opening up the planning process to a wide range of voices. But the implementation of policy requires that this opening up process, sometimes called 'deliberative' or 'discursive' planning, engages with the formal institutions and agencies of planning. The second investigation focused on the barrier to implementing new transport policy resulting from deficiencies in the horizontal and vertical integration of policy – between government agencies at the same level (State or local), and between levels of government.

Understanding the barriers to change is a necessary first step in a transformative programme. In the next section of the book we move on to strategies to promote change.

Strategies for local change

The theory of democratic politics is supposed to contain the means of change. Democratic constitutions provide for periodic recall of the government of the state so that the public can exert influence. Pluralist theorists argue that interest groups or pressure groups exert influence directly on government, and policy responds according to some kind of compromise among the policy directions demanded by different groups (Waste, 1987). Influenced by elite theory and Marxism, neo-pluralists posit that some pressure groups are 'more equal than others' – to use Orwell's phrase (see Lindblom, 1977). Because business – or capital – controls investment, governments ultimately have to succumb to its demands. Neo-liberals, on the other hand, argue that businesses are subject to the rule of the market, which represents the interests of consumers, and governments should only be influenced by periodic democratic voting under the constitution (Buchanan and Tullock, 1965). Pressure groups representing special interests, NIMBYs, merely divert governments from acting in the public interest represented by the market and democratic voting.

All of this classic 'assenting' theory of liberal democracy assumes that the public controls government, and government controls the state (see Low, 1991). Modern political theory, however, offers a more complex view of governance and social change. The relationship between government and the state is not a simple matter of control and accountability. States are not monolithic hierarchical structures but infused with networks of power (see Rhodes and Marsh, 1992; Considine, 2005). The key to understanding strategic change lies in understanding the governance of transport in cities, and this is the subject matter of Chapter 7.

The cities of the East and South demonstrate extremes of social geography, with the very high density, high rise cities of, for instance, Singapore and Shanghai, and the low density, low rise cities of Australia and New Zealand. Governance systems vary widely, offering different barriers to and opportunities for sustainable transport. Australian and New Zealand urbanism is most often discussed in relation either to European or North American models. But increasingly, delegations from China and other Asian countries are visiting Australia and asking what they can learn from the experience of Australian transport planning. With an urban population expected to grow rapidly over the next twenty years – resulting from Australia's mining boom feeding demand from China – Australians are also asking what can be learned from the experience of high density Asian cities.

In Chapter 7, we compare examples from Singapore and Shanghai on the one hand, and Melbourne and Perth on the other. In keeping with the integrative philosophy of the book, we find that comparing *unlike* examples leads to new insights. With a common aim – that of reducing carbon emissions from transport – policies have to be designed that are fitted to widely varying local social and governmental circumstances. Global solutions, such as increasing density to reduce fossil fuel consumption (as initially derived from the work of Newman and Kenworthy, 1989) have been seriously challenged (Troy, 1996; Mees, 2010). Yet it is clear that the relationship between land use patterns and transport systems has a significant role to play in moving towards sustainable transport, requiring the integrated management of these two policy domains in 'transit oriented development' (Curtis, Renne and Bertolini, 2009).

The chapter begins with an analysis of the social geography and transport governance of the four cities. Comparisons are made between the Singapore experience (Han, 2009; Yang and Lew, 2009) and that of Shanghai (Pan 2011), and the different approaches to land use and transport policy in Melbourne and Perth (Stone, 2009b). The second part discusses the transport policies being pursued in the cities, and the path dependencies, problems and issues that are arising, as well as the solutions being implemented. The final part brings together geographies, policies and governance structures and considers the lessons that the low density and high density cities have for each other in terms of transport governance.

How then does change occur of the kind and scale necessary to resolve the mobility dilemma? We address this question in Chapter 8. The answer can at best be only sketchy and partial. Deep or 'third order' change is 'marked by radical changes in the overarching terms of policy discourse associated with a paradigm shift' (Dudley and Richardson, 2000: 18; see also Hall, 1993). Such change is a 'disjunctive process associated with periodic discontinuities in policy' (ibid.).

In this chapter we show that, common to all stories of deep change is the consistent application of policies to support alternatives to car-dependence over periods of a decade or more: other influences such as wealth or cultural attitudes to mobility appear to be less important determinants of change – or of failure to change. The key to this process is the development of particular programmes and institutions that are appropriate to the local context, and which, through their emerging success, extend the constituency of support for the model of sustainable transport as described in Chapter 4.

Theoretical understandings show how local politics can be an important mediator of apparently irresistible global economic forces (Leo, 1994). However, to commence the process of change from the dominant 'car-based' conception of transport policy within government and professional agencies, considerable political and social momentum is required. This is not simply a question of demonstrating the rationality of the alternative – although this is also important work as will be discussed in the following chapter. Mutually exclusive, competing conceptions of transport policy are contested in the local political arena. The tight institutional control exercised by established policy networks over problem definition and 'solution sets' must be challenged. This appears to be possible only during local 'windows of opportunity' (Dowding, 1995).

This chapter discusses the pattern of political and professional action in six cities in Europe (Bratzel, 1999); Melbourne and Perth in Australia and Vancouver, Canada (Stone, 2008; Legacy, 2010a and b); and Portland in the USA (Leo, 1998). This pattern includes:

- serious challenge to the government of the city or region due to popular opposition to elements of its transport and environmental policies: the 'window of opportunity'
- change in political leadership and in important transport policy positions
- long-term institutionalization of operational and technical changes through the creation of a new policy network and extension of the mandate for new policies among an increasing range of social actors.

In summary, then, strategies need to begin with a strong discourse coalition in the community which is then able to mobilize change within government either by inserting its leaders into government positions or by finding sympathetic champions within government. This coalition needs to be sustained, and must be able to shape a new discourse capable of gaining public support. Critical to successful implementation of change is mobilizing the apparatus of the state at an early stage. The time scale for this process is long, so advocates of change have to prepare for a long and continuous struggle for acceptance. Timing of interventions is likely to be crucial so that new ideas can find a place in discursive foundations that have already been prepared.

In Chapter 9, we turn to the knowledge base of transport planning. We explain in this chapter a new set of practice technologies, developed within the GAMUT research project, which can assist planners and decision makers when confronted with complex transport problems as they move towards sustainable transport. In recent years, a plethora of new approaches to transport planning has emerged that take account of environmental and equity concerns, and with a new emphasis on public transport (Geurs and van Eck, 2003; Preston and Rajé, 2007; Dodson and Sipe, 2007). These are discussed briefly in this chapter. Three new decision tools associated with the GAMUT research are reported in greater detail. The first is a spatial network analysis tool bringing together 'place' and 'node' features of transport systems (Spatial Network Analysis for Multimodal Urban Transport Systems, SNAMUTS). The second is focused on supplementing conventional four-step transport modelling with additional approaches in order to deliver greater awareness of and insight into the potential impacts of specific transport and land use scenarios (Modular Urban Land Use and Transport Tool, MULUTT). The third applies European 'network' planning for public transport in the context of Australasian cities, developing a benchmarking tool to support planning of new public transport service patterns (Nielsen et al., 2005).

As already noted, change of the assumptions about mobility and transport needs to occur throughout the community, starting with children and their parents or guardians. This is the subject of Chapter 10. Children are the transport decision makers of the future. The decline in children's autonomous use of public space has had negative consequences, for their health and also for community wellbeing. Conversely, taking a child's-eye view in making our communities safe, accessible, and welcoming may well improve outcomes for all.

The dilemma of mobility discussed in the first part of this book has hit children particularly hard. The definition of children's independent mobility (CIM) is the ability of those under 18 to explore their communities autonomously, without adult supervision. The erosion of children's independent mobility directly contributes to the abandonment of streets and public spaces through their absence. It indirectly contributes to car dependence through their parents chauffeuring their children to home, play, and friends' houses, leaving children virtually trapped in their homes and back yards by the vicious circle of more cars leading to less safety for pedestrians, leading to increased fear.

This chapter reports a research project aimed at finding ways of breaking out of this vicious circle. The research identified interventions that facilitate children's independent mobility within urban environments, and arrived at principles across interventions that can be replicated in different settings. The emphasis was on determining the role of both physical and social planning to assure children's 'right to the city' and its relationship to improving health, citizenship, and sustainable transportation outcomes. While the geographic focus was Victoria, Australia, we drew on a wide range of Australian and international research and policy literature, particularly in the search for best policies and practices.

These new ways of thinking about transport discussed in the preceding chapters, and the new paradigm of sustainable transport, present a problem of how best to disseminate this knowledge in the wider community. This is the topic of Chapter 11. The transformation envisaged and promoted by the GAMUT research project from automobility to sustainable transport is a political act embodying new sets of values. Thus the generation and dissemination of knowledge flowing from the project is not politically neutral. It goes to the heart of real interests within the fields of government, transportation and infrastructure. This reality has to be recognized in any process of engagement between the academic and broadly political domain – that is the domain in which these values and interests play out.

The role of notionally independent bodies such as universities has been steadily eroded. Increasingly, in Australia, universities have experienced reductions in freedom and untied funding. The expansion of contract work for government and private bodies has had the foreseeable effect of reducing the impact of excellent research done by such bodies, as a perception of bias grows among members of the broader community. Contributing to this perception is also the increasingly vocal public commentary by individuals and groups within universities supporting particular positions – for instance on climate change and transport – essential though that public engagement is to informed public debate.

A dialogic strategy is therefore needed to raise the level of engagement of the informed and interested public, and to increase participation to raise the credibility of the knowledge generated by research (see Torgerson, 1999; Eckersley, 2004: 164). The process is based on the understanding that there will be an *exchange* of knowledge rather than a one-way flow (long urged by, for example, Lindblom and Cohen, 1979). This strategy is guided by the need to build and strengthen the epistemic community around sustainable transport (see for instance Haas, 1989). The focus of such a community is the transport and land use planning aspects of the response to environmental change associated with transport systems – necessarily within a global perspective. As Haas observes, epistemic communities are 'networks of knowledge based communities that are both politically aware, empowered through their claims to exercise authoritative knowledge, and motivated by shared causal and principled beliefs'. This concept parallels the position of engaged social science (or *phronesis* following Flyvbjerg, 2001). This chapter describes examples of the practical approaches undertaken by the GAMUT research team in Australia to build the epistemic community through dialogue, via forums focused on specific issues that have engaged a wide range of government and community stakeholders.

The final chapter (Chapter 12) uses four broad ideas to express the synthesis of the book. First the chapter shows how the different conceptual and theoretical approaches are woven together and related in order to reflect on the dilemma of mobility. For instance, turning points, critical junctures, discontinuities, or points of transformation are found within different theories of social change and stability. Within the framework of practical knowledge, or phronesis postulated by Flyvbjerg, insights are gained from geographical and environmental analysis, and investigation of governance based on theories of regulation, path dependence, governmentality, and epistemic community. Second, the chapter reviews the different levels and domains in which the mobility dilemma and transport problem has been understood and investigated in this book: at global and city levels, and within technical, institutional and cultural domains. Third, the book reviews the different experiences revealed of governance and management of transport in the East Asia-Pacific region and areas of potential common learning. Finally the chapter brings together the solutions proposed to the dilemma of mobility, suggests lessons for future governance of transport in different contexts, and reflects on unanswered questions.

Conclusion: transforming knowing

The book's epistemological position is that of social science. That is to say the knowledge, such as it is, of the way human societies behave. But, of course, human societies behave as systems responding to their understanding of an external environment, as well as to the technological possibilities they create through the knowledge supplied by the natural sciences. Our aim here is to bring to bear upon the questions of mobility and transport what is known about their human impact on the environment. In this respect there is no absolute certainty but only probability. Human society interprets scientific probability in terms of risk: how likely is the event to occur, and how serious are the consequences if it does?

Compared with natural science, social science is still in its infancy. Theories abound, but there is not the sustained and sophisticated effort to test them empirically that provides the driving force for the development of knowledge in the natural sciences. We agree with Flyvbjerg (2001) when he proposes that social science cannot emulate the detachment of natural science and should not try to do so. Humans studying the human societies of which they are part interfere fundamentally with observation so that objectivity is impossible. Whether that is true for all time or just at the current stage of development of social knowledge remains an open question. But, for now, with social science in its present state of development, it seems prudent to view this region of knowledge, as Flyvbjerg (2001) does, in terms of *phronesis* translated from Aristotle's *Ethics* as practical common sense.

Unlike *episteme* in which values have no place, *phronesis* fuses knowledge and values in an interplay of what is and what should be. If we want to plan a social artifact such as a transport system, then we have to know both where the system is heading, and portray where it ought to be heading. For instance, the concept of 'mobility dilemma' both values mobility and projects its dangers. Flyvbjerg defines phronesis as 'ethics, deliberation about values with reference to praxis; pragmatic, variable, context-dependent; oriented toward action' (ibid.: 57). The point of departure of this book can be stated precisely in terms of the three questions Flyvbjerg (p. 60) posits as the core of social science understood as phronesis:

1 Where are we going?
2 Is this desirable?
3 What should be done?

Where are we going? Our analysis tells us that the world is heading for environmental catastrophe and the transport modes pursued via automobility are complicit in this trajectory. Is this desirable? Well, present human desire certainly drives automobility so in one sense, yes, it is desirable. Yet for those whose lives are ended or horribly altered by crashes and pollution, for human populations of the future, for other species on earth, and for those who do not benefit from automobility, it is far from desirable. What should be done? We have presented the solution in terms of the change of paradigm from *automobility* to *sustainable transport*, and some ways of making the change. But, let's face it, we know so little about how human behaviour and human institutions change – and today the world faces a problem of global scope without adequate institutions of global governance. This book, we hope, joins others in adding to the fund of knowledge attempting to answer that crucial, final question.

Note

1 This is an actual not a hypothetical example; the names are fictitious for reasons of confidentiality.

2 The mobility of goods and people

Kevin O'Connor

Introduction

One of the predominant features of our current economy and society is the level of mobility of its people, goods and information. Economic, social and cultural patterns are intimately linked to this mobility, which can be seen in the location and type of transport infrastructure, and patterns of economic activity, housing and associated social and community facilities in cities. This chapter explores this perspective via an emphasis on the way the economic dimensions of transport are expressed in city development. It illustrates that mobility, facilitated by a diverse transport system, is an essential part of our economic context. In turn that means that plans to transform transport by changing mobility by mode, by distance travelled and/or frequency of movement, need to recognize this economic context and the infrastructure and transport services it entails.

The global mobility of goods: sea and air freight

Sea transport has a very long history in the mobility of humans. As sailors ventured further and further into the unknown from Europe and also from China, new maps of the world appeared, and new locations were connected by trade and later by the migration of people. In much of this long history, the role of shipping was to move raw materials from the newly discovered peripheral colonies to the core industrial economies of Europe, and return with finished goods, and with people. Very recently these patterns have changed. Much industrial production is now located in what were once peripheral locations creating what Dickin (1992) described as a 'global shift' in economic activity. This shift means the globe is now crossed by container vessels and air freighters moving components to and from the new production sites and carrying finished goods to consumers. What are now labelled 'global production systems' are central to the day-to-day operation of the global economy, and their products can be seen on the shelves of retail outlets in most cities around the world. In fact, a greater variety of goods are carried more frequently over longer distances than ever before.

Rodrigue et al. (2009) have shown that the value of trade has risen to be around half the value of world gross domestic product (GDP), whereas it was around 20 per cent in the early 1980s. That rise has been matched by a steady increase in the value of all exports. Taken together these measures mean that mobility of goods is now proportionately twice as important as it was in 1980, showing that mobility is a critical element in the way firms now produce goods.

Urban impacts of the mobility of goods

In the past, sea port activities provided a very significant focus for transport activities, initially in a cluster of narrow roads and lanes in the port district, later complicated by a grid of rail lines and roads. Those outcomes have been re-expressed in the current context as the global mobility of goods has been channelled through a small number of locations. Just as large ports began to dominate smaller ports in the early era of shipping, so the adoption of the container, and the hubbing of air freight, continued the trend. To an extent, this reflected the need for specialized infrastructure, but it also reflected the fact that the new large ships and specialized air freighters needed large and regular loads – and bigger port and airport facilities – which could be gathered together at just a few places (Hayuth, 1981).

To illustrate that effect, research shows that 73 cities ranked as 'Worldwide Centres of Commerce' by Mastercard Worldwide (2008) account for 64 per cent of all air freight and 58 per cent of all sea freight moved through almost 1500 sea and airports for which data was available in 2006 (Table 2.1). Recognizing the role played by another 20 important freight cities, the table shows that around 80 per cent of all sea and air freight moves through just 93 cities around the globe. Finer scale analysis found just eight cities, which had multiple sea and airports, were responsible for 25 per cent of the world's sea and air freight movements, a share that has risen steadily since 1996 (O'Connor, 2010).

This concentrated outcome has some serious implications for transport operations and planning. On the operations side, the physical infrastructures of the sea and airports occupy large land areas, and the connections to regional and national markets require high capacity road and/or rail routes, themselves requiring additional land along corridors. Yet the cities with large shares of traffic have among the most expensive commercial and industrial land in the world, so there is pressure to find lower cost operating sites within and around these cities.

That pressure, allied to the needs of modern supply chain management operations, has led to low-density land development for these activities. As an example, market research in Europe has found that, 'distribution centres may be located 60–70 kilometres way from ports, if sufficient transport connectivity is in place' (Angel et al., 2006: 29). This outcome has been labelled 'freight sprawl' in the New York Metropolitan Region (de Cerreño

Table 2.1 Location of logistics activity among cities of the world

Type of City	*Number of places*	*Share of global air freight (tonnes lifted) (%)*	*Share of global sea freight (containers moved) (%)*
Global city* with sea and air ports	44	49	59
Global City with airport only	29	15	
Total for global cities	73	64	59
Other major transport cities: (Top 20 ranked on share of sea or air traffic)	20	18	19
Major sea and air transport cities	93	82	78
Rest of cities in databases (859 airports, 437 seaports)		18	22
Total		100	100

Note
* As defined by Mastercard Worldwide (2008)

et al., 2008: 3). It is also seen in a vast expanse of new warehouse space in the 'Inland Empire' east of Los Angeles (United States Government, Federal Highway Administration 2009) and regional networks surrounding ports in north-west Europe (Notteboom and Rodrigue, 2005). The location of warehouses and storage space – along with the rapid expansion of manufacturing – has also been acknowledged as a key influence on the development of Asia's mega cities (Webster, 2002).

This outcome has a major consequence for the transport network as trucks are used to move freight between port and airport and the storage centres or the factories. Truck congestion and truck queues are a common feature of the major port cities of the world. A further effect of these patterns is the final delivery stage from storage locations to final users. These delivery routines, linking storage centres to dispersed consumption points, create local networks of trucks, delivery vans and courier vehicles that contribute to around-the-clock traffic patterns across the breadth of the metropolitan area.

To understand these outcomes it is necessary to recognize that, 'a new distribution sector is emerging, no longer derived from retail or manufacturing, but increasingly following its own logic, unfolding its own power … ' (Hesse, 2004: 1036). The operation and planning of this sector contributes to the spread of jobs and housing. Workers access these destinations by car as the new logistics facilities and factories are rarely served by the suburban commuter rail system. Some will use the arterial roads and freeway systems built in part to enable the movement of freight. The local pressures associated with truck transport lead to calls for additional road construction, or the designation of truck lanes on certain roads. Efforts to move more freight by rail create conflicts in the use of local and regional rail systems so that dedicated freight tracks might be provided. Any expenditure on related transport infrastructure further consolidates the role of the port city in national and regional transport networks, and increases traffic to the port and airport.

Hence the high mobility of goods in today's global economy has shaped the geographic spread of metropolitan areas. Any attempt to transform transport will need to recognize the capacity, opportunity and problems of this infrastructure and its traffic flows, and its potential for further change in scale and location.

The global mobility of people: air transport

Perhaps no other mode of movement illustrates the rise in frequency, distance and participation in mobility across social class than air transport. Air transport was little more than a defence activity until 1945. It then emerged with infrequent services for the wealthy and for business people, and involved short journeys initially within nations. It has expanded over just 50 years to a mass market activity, over both long and short distances with very predictable, usually daily – or on some national routes, hourly – frequencies. Rodrigue et al. (2009) have shown the rate of growth in air traffic has been faster than the expansion in world GDP. This outcome reflects the simultaneous development of aircraft and navigational technologies that have allowed large aircraft to fly long distances at lower cost between major markets (Winchester, 1990).

Passenger air transport has grown in response to demand for people to visit friends and family, for tourism and for business. The social and community element of air travel usually mirrors earlier migration movements: links between old colonial nations and their European origins are a strong element in current global air passenger movements. Tourist traffic focuses on the character of destinations, and the big cities of the world are major destinations. The direction of business travel has been shaped by the global production systems discussed above. The geographic separation of financial services, management, design, component production and final assembly of a good or service requires the mobility of people engaged in these activities.

Once again one might expect a dispersal of these disparate air passenger movements to all parts of the globe, but in reality there is a strong concentration of air traffic in a few of the world's major commercial centres, as Smith and Timberlake (2002) and Zook and Brunn (2006) have shown. This geographic outcome means that a few cities have the world's busiest airports, and hence that local transport outcomes in these regions feel substantial pressure.

The urban impact of airports

Airports have emerged as important centres within metropolitan areas as they have several layers of activity. One layer is the basic transport function, including the technical aviation services and the organization of passenger movement, including customs and immigration. In addition, there is ancillary activity represented by retail stores, hotels, car hire agencies as well as the air freight-related storage and warehousing activity. On a third layer are the business activities located in nearby business parks. Airports have become attractive sites for business in recent years as they provide national and global access. In particular, it has been shown that airports of metropolitan areas are a magnet for high-tech businesses (Button and Stough, 1999). This outcome can be seen in a number of places, notably around Amsterdam, Heathrow, Chicago and Los Angeles. As an example, Brueckner (2003: 1466) has shown that a 25 per cent increase in passenger traffic through an expanded Chicago Airport would be associated with an additional 90,000 jobs.

The development of air passenger movements has created problems in many cities. These have been addressed in different ways and, through these responses, airports have contributed to re-shaping the local and regional transport infrastructure. The first approach was to intensify the use of existing facilities. This can be seen in the continued development of airport capacity at Heathrow, Amsterdam, Tokyo (Haneda) and Chicago. By emphasizing the existing facility, these cities have strengthened that facility's impact upon the local transport network, so justifying additional road and rail links.

A second response to the increase in air passenger traffic – in some cases a recognition that capacity limits have been reached at the original facility – is to make use of other airports within a 100–150 kilometre radius, creating a multi-airport network (Derudder et al., 2010). In London, although Heathrow has been expanded, Stansted, Luton and Gatwick have become important international airports. The same outcome can be seen in San Francisco – Oakland and San Jose provide two additional airports – and Washington – with Baltimore–Washington added to the two other airports in the region. New York has long had three major airports and Paris, Tokyo, Seoul, Shanghai, Bangkok, Osaka and Milan now have two. Some of these secondary airports have international connections, others provide national links, and some provide bases for low cost carriers. If traffic volumes are high enough, rail links are provided – as is the case for Gatwick, and Paris Charles de Gaulle; often, however, these smaller airports rely upon road links, utilizing the existing – or in some cases upgraded – links in the metropolitan freeway system. With this enhanced transport infrastructure these airports provide another set of accessible work opportunities, so adding to the complexity of the patterns of daily movement across the metropolitan region.

In some cities the original airports were too small and too difficult to extend, and new airports were built. Good examples are Denver (Dempsey et al., 1997), Tokyo (Narita), Hong Kong and Bangkok. In these cases the new facility was built in an urban fringe location, providing a new employment and transport node in what was at the time a distant location. Provision of new road and rail links often has the effect of creating a new corridor of urban development stretching out to the new airport.

Hence the volumes of passengers moving through airports alone make them one of the major nodes in the metropolitan area transport system. That role is reinforced by the business activity located in the area surrounding the airport, which tends to be larger in the busiest airports. Coping with those twin demands has triggered investment in metro and regional rail – and, in Europe high speed inter city rail – as well as local and regional road connections. In many cities, the road connections are part of a ring road or beltway network, contributing to circumferential travel patterns. These infrastructure and service responses have reshaped the day-to-day movement within the metropolitan area with the airport corridor often being one of the busiest in the metropolitan area. As with the movement of goods discussed above, the airport-focused infrastructure and its daily flows need to be seen as important considerations in efforts to transform the transport system.

Metropolitan mobility of people: the complexity of commuting

Some of the infrastructure and related activities of sea and air transport outlined above have intersected with five significant influences that emerged over the past 30–40 years to cement a new structure for metropolitan transport. These forces were the growth of cities; the shift from public to private transport; the increased role of women in the workforce; the growth of manufacturing and services in suburban areas; and the expansion of the housing supply. The combined outcome of these changes has usually been labelled 'urban sprawl' and is often associated with US urban development, but recent research in Europe (RWI et al., 2010: 66) shows 'the outer zones of all different city types on average have been growing faster than the core zones'.

For transport planning these changes have meant large increases in the numbers of people moving across a different spatial pattern. Studies of suburban development showed that it was not simply sprawl, with more trips to the core of the metropolitan area, but rather it involved a new set of connections between the locations of housing and jobs. Those new connections create a set of opportunities and also constraints on efforts to transform transport.

The transport dimensions of the suburban economy

Analysis of the scale and direction of this new outcome usually begins with the changes in population and population growth, and the common outcome is that outer areas have expanded and inner areas have declined in relative terms. Even though, in recent years, inner area populations have grown rapidly in many cities, outer area growth is still significant, and the majority of the population still resides in outer areas. Population is, of course, just one element of urban change, and for transport planning may be less important than the influence exerted by the location of employment. Trends in the location of employment have shown high levels of suburban growth. In US metropolitan areas, Glaeser and Kahn (2001) show that suburban areas now account for around 70 per cent of all US metropolitan area employment. In German cities the figure in 2006 was 66 per cent (RWI, 2010: 66), while Gaschet's (2002) work on French cities showed that, in the case of Paris, 62 per cent of jobs were outside the central city. In Australia, O'Connor et al. (2001) showed that between 65 and 70 per cent of jobs in Australian cities are in the suburbs; detailed research by Davies (2009) showed the share of jobs located in the suburbs rose from 65 per cent in 1991 to 72 per cent in 2006. Recent developments in Asian cities indicate that the outer suburban fringe is attracting the majority of new manufacturing investment, so tipping the distribution of employment in favour of these locations (Webster, 2002; Wu and Phelps, 2008; Alpkokin et al., 2007).

The growth in suburban employment is associated with a steady increase in suburb-to-suburb commuting. Pisarski's (2006) analysis of commuting in the US shows how this outcome has emerged. Using national data he showed that in 2000 there were 40.8 million suburb-to-suburb work trips, more than double the 16.6 million who travelled from the suburbs to the central city. It was also a much greater number than the 24.5 million who travelled around the central city. He observes, 'commuting from suburbs-to suburbs accounts for 46 per cent of commuting activity (including the flow to suburbs in other metropolitan areas) with the "traditional" commute from suburb to central city at 19 per cent' (Pisarski, 2006: 47). Comparing this situation to 1960, his analysis shows, 'suburbs-to-suburbs commuting has quadrupled in the period while central city flows have stabilised' (p. 52). This feature of travel reflects the fact that many households match suburban jobs with their suburban housing, which in turn has had the effect of maintaining average travel times over a 30-year time period in the Washington DC metropolitan area (Levinson and Kumar, 1994).

Although these outcomes have usually been associated with cities of the US (Lee at al., 2006), the strength of intra-suburban travel is apparent in cities as diverse as London (Spence and Frost, 1995; de Goei et al., 2010), Madrid (Garcia-Palomares, 2010), Paris, Lyon and Marseille (Aguiléra, 2009), Copenhagen (Hansen and Winther, 2007) and Melbourne (O'Connor and Healy, 2004). These cross-regional trips were almost always made by car, so the steady expansion in the importance of suburban employment and housing played a large part in the change from public to private transport mode observed in most cities.

Over time, a geographic structure could be seen in this outcome in some metropolitan areas as employment growth occurred in suburban centres, seen in research on Los Angeles (Giuliano and Small, 1999) and Chicago (McMillen and McDonald, 1998). This outcome has been observed in cities in Germany (Guth et al., 2009) and in France (Gaschet, 2002), as well as cities such as Dublin (Vega and Reynolds-Feighan, 2008), Madrid (Garcia-Palomares, 2010), and Brussels (Riguelle et al., 2007). In some US cities very large concentrations of employment emerged, usually at or near beltway junctions, and were labelled 'edge cities' (Garreau, 1991). At a smaller scale, Kotkin (2005) has identified 'suburban villages' in recent urban change in the suburban parts of US metropolitan areas. A variant of edge cities has been found in European conurbations (Bontje and Burdack, 2005; Phelps and Parsons, 2003) although this phenomenon is not so evident in Australian cities (Freestone and Murphy, 1998). Where clustering of jobs in outer suburbs occurs it is potentially important for transport planning as such clusters represent a focus of demand that could be served by local and regional public transport services. However, that potential is limited by the fact that, in many cities, research on suburban centres finds that there are more jobs outside the centres than within them.

The new suburban economy emerged in metropolitan areas where the public transport system was designed to move large numbers of people along tram and train commuter corridors to workplaces in the city. That meant that getting to a suburban job required a car. Initially arterial roads were expanded, and centrally focused freeways provided travel opportunities for outer suburban residents with mid-suburban jobs. Eventually the efforts of the road builders shifted to the movement of freight between production sites and suburban logistics centres, requiring beltways, ring roads and périphérique routes. These are now the distinctive feature of road transport in modern metropolitan areas and have a very significant impact on accessibility to population and employment, as Gutiérrez and Gómez (1999) have shown using data on the M40 around Madrid. These roads are often the most heavily used in each metropolitan area.

Though research tends to emphasize commuting, residents of suburban areas also travel for business connections, social life, education, family ties and entertainment. Pisarski (2002) estimated that the work trip in fact accounted for:

> roughly 20–25 per cent of local travel. While commuting has grown rapidly in the last 20 years, trips for personal business, shopping, etc. have grown even faster. Total trip-making per household has grown 66 per cent since 1970 despite a 17 per cent decline in household size.

In effect there is a deep layer of complexity in the links between suburban residences, jobs and services. Understanding that complexity means following the research of Bruegmann (2006) to move beyond long-established images of 'urban sprawl' that portray the main problem as the long commute of the urban fringe resident. Rather, the challenge of the complex mobility of suburban residents is how public transport can be re-designed and re-programmed to provide a substitute for the private car, and perhaps how the private car itself can be changed, especially given the expected scarcity of oil (Dodson and Sipe, 2007).

It is also important to recognize that, although suburban employment and housing development has become the predominant form in most cities, the role of the CBD and its surrounding inner city area remains significant. That significance is expressed in a continued role in advanced services, government activities, education and tourism-related sectors. Indeed, in some cities, the clustering of high level services maintains specialized districts developed around a given skill, seen in areas specializing in advertising, legal services, government and finance. The vitality of these activities has been matched by an expansion in residential development, often using land and buildings from earlier manufacturing, rail-yard and seaport functions. New residents in these areas are sometimes able to rely on the established public transport network to reach inner city work destinations, which has contributed in part to the increased use of public transit in many cities. That set of circumstances has maintained the earlier problems of congestion in and around the inner city of most metropolitan areas.

Mobility in the modern metropolis: challenges for transport

The long-established centrally-focused transport grid of rail and tram was matched to a centrally-focused distribution of people and jobs in a pattern that has been labelled the mono-centric city. In many cities this structure fitted local planning approaches as the geography of the transport services was closely aligned to the geography of an administrative organization. 'London Transport' was a good example of that situation, providing multimodal services within the Greater London Council area.

New spatial patterns

Shifts in the mobility of people and goods associated with the new business, community and social organization outlined above have changed the old situation in two important ways. The first is in the scale of the area, and the second is its internal structure. The first change is expressed in new descriptions of cities, which include terms such as 'One Hundred Mile City' (Sudjic, 1992), 'Extended Metropolitan Region' (McGee, 1994), 'Global City Region' (Scott, 2001) and 'Mega Region' (Florida et al., 2008). These terms share notions that cities are now far flung, that the logistics infrastructure of sea and air ports are significant elements, and the parts are connected by a network of roads with prominent circumferential routes – sometimes with a separate inner and an outer ring road. Examples have been described in the UK (Breheny, 1999), the US and Canada (Florida et al, 2008), Europe (Knapp and Schmidt, 2003) and Australia (O'Connor et al., 2001) and are seen in the spread of urban regions around cities in Italy, (Dematteis and Governa, 1999), Sweden (Warneryd, 1999), and

France (Bretagnolle et al., 2002). Webster (2004) shows how this new scale of development applies to the growth of Bangkok and also to Shanghai (Webster and Muller, 2002).

The second change is a new internal structure, usually described as a 'polycentric city' (Kloosterman and Musterd, 2001; Hall and Pain, 2006). The essence of this idea is that there are now a number of centres of varying size that serve the business, social and community needs of populations spread across a large urban region. One of these will be the original core or CBD of the city. Although descriptions are imprecise, as it is not easy to determine the number of centres and the strength of their connectivity (Parr, 2008), this image has become a broadly accepted view of the way metropolitan regions now work on a day-to-day basis.

Transforming transport in polycentric metropolitan areas

The polycentric form offers both potential and problems for efforts to transform transport. The potential lies in the scope created to meet local as well as regional mobility needs. It is possible, too, that shorter trips will be available to some residents, bringing jobs or community facilities within the range of cycling or even walking, and eventually of electric vehicles. The dispersal to centres and across sub-regions could also reduce the congestion around older core destinations. Intra-suburban car trips will use less fuel than the longer commute to the central city, while the energy used on the heavy rail suburban systems could be reduced. However, the polycentric form has transport problems. There are now many origins and many destinations spread at low density rather than clustered at nodes. Travel volumes on individual routes may not be sufficient to justify public transport. For the rail system, the far-flung nature of the metropolitan area will generate conflict between long-distance higher-speed rail connections and slower local services, possibly requiring additional track alignment. As private road transport is likely to remain a major aspect of transport in these new polycentric far-flung regions, new road construction may still be needed.

Notwithstanding these problems, a number of cities have adopted the polycentric framework as a new starting point for transport planning. For example, Casello (2007) has carried out a modelling exercise to show how a polycentric transit system could be fitted to Philadelphia. He finds that significant gains in modal split in favour of public transit can only be achieved with simultaneous service improvements and restrictions on car use. He also notes that services have to both connect centres as well as allow movement around centres.

Planning for transport developments in London incorporates the polycentric insight by adopting a set of sub-regions as the planning framework (Mayor of London, 2004a). Policy thinking then explores ways that infrastructure development and service provision can address the needs of these regions. One element in a massive policy effort is designed to:

> ... strengthen orbital public transport and support spatial development priorities. [These approaches] ... will relieve public transport in central London by facilitating access from one radial route to another without the need to travel via the Central Activities Zone; they will improve access to a number of town centres and suburban areas and potential development areas; and they will support a number of priority regeneration areas.
>
> Mayor of London, 2004b: 143

The focus on improving orbital travel is designed to deal directly with local demand for short trips, while also providing regional connectivity to major nodes. This approach lies behind a proposed light rail service across a section of suburban Copenhagen (Andersen et al., 2006) and a large system of cross radial lines proposed for Paris (The Transport Politic, 2010). Orbital bus services have recently been introduced to Melbourne, Australia.

In the longer term, realizing the potential of the polycentric form for transport change will require changes in services as well as control over the spatial arrangement of new activities, as Curtis (2006) has outlined in Perth. This approach will draw upon ideas developed for transit-oriented development (Curtis, Renne and Bertolini, 2009). That involves land use management policy, and action to ensure that new employment activities, significant community services and some residential developments are placed in locations that are served by cross-regional transport services. The full benefit of this approach requires a well-designed world where a hierarchical system of centres can be paired to a hierarchical arrangement of transport services such as Morichi (2005) suggests will be needed in rail systems in Asia's growing mega cities. In most metropolitan areas, however, the spatial legacy of the past centres and sub-regions will remain an important influence on the planning approaches.

That is especially obvious in planning for the movement of freight across a polycentric region. This activity creates nodes – the airport and seaport; corridors – the road and rail links; and destinations – the storage and distribution facilities – and might require dedicated infrastructure – separate rail lines and road lanes. Approaches to manage this element of the transport system are becoming increasingly common in US cities, as outlined by the research of the Federal Highway Administration (United States Government, 2010), and are exemplified in the Alameda Rail Corridor between the ports of Long Beach and Los Angeles and the rail terminals near the centre of that metropolitan area (Yang, 2008; Giuliano and O'Brien, 2009; Hauswirth et al., 2004). This provides a dedicated rail track for container movement which replaced an old, slow above-ground route.

In most metropolitan areas the freight transport nodes, corridors and destinations will be part of the overall transport system, with trucks and commuter traffic mixed on the roads, and passenger and freight services mixed on the rail network. This can be seen in the 'Gateway Project' in

Vancouver (Government of British Columbia 2003, Ginnell et al., 2008). The approach acknowledges:

> Metro Vancouver has also become an important gateway for international trade and has emerged as Canada's hub for Asia-Pacific trade. Over the last ten years, the Port Metro Vancouver has seen increased trade with China and other countries and that growth is expected to continue over the coming years.
>
> Government of British Columbia, 2003, no page numbers

The project has four spatial foci across the metropolitan region and includes road widening, new road construction, new bridges, and upgrading interchanges. Although this construction is justified mainly by the freight movements, it is also expected to provide scope for commuter travel. So one particular project is:

> a new four-lane, 80 kilometre per hour route along the south side of the Fraser River ... (to) ... provide a continuous and efficient route to serve the port facilities, rail yards and industrial areas along this key economic corridor, and will also benefit commuters.
>
> Ibid.

These examples show that incorporating the modern polycentric structure into efforts to transform transport will mean redesigning the types of services provided – new small-vehicle local networks, some long-distance express links – and the selection of destinations based upon job density and community services rather than simply selecting nodes on the original rail system. The approaches will also need to recognize that the car and truck will remain a significant part of transport services in the polycentric metropolitan area. Hence the transport planner will need to consider expansion in park-and-ride facilities that connect the car trip to a local or regional connection, and freight centres that could reduce truck and courier trips by concentrating the first part of the delivery task on larger scale regional services as outlined by Slack (1999). Finally, transport planning will need to be engaged in the application of policies that encourage the use of small fuel-efficient engines and LPG fuel by applying lower registration fees and lower speed limits, for example, as well as supporting the adoption of electric car technology linked to renewable static energy sources.

Governance is a special issue in drawing out the potential of the polycentric networks of jobs, infrastructure and residences to help transform transport. The difficulty of creating a coordinated strategic approach in a diverse array of local communities is a well-worn path in regional strategic planning (Salet et al., 2003). The polycentric form adds to the complexity of that path by adding more communities to the mix, while transport projects themselves draw in a broad array of actors and interests. Hall and Clark (2010) have shown how competing layers of national government, regional authorities, local communities and logistics companies have delayed and reshaped the investment in the Gateway project in Vancouver outlined above. In contrast, Callahan (2007) provides a positive analysis of those issues with respect to transport infrastructure planning in Los Angeles, showing the significance of the local context and the potential for cooperation that can flow from new resources created in large projects.

Conclusion

The increased mobility of goods and people has had a major effect on the structure and organization of most metropolitan areas. That structure has a deeply laid foundation expressed in the distribution of economic activity, community services and population, and that foundation will need to be acknowledged as future transport solutions are developed. That foundation offers major potential to transform transport, allowing the mobility of goods and people to be expressed via some new nodes, and through the proximity of jobs and community facilities to residential areas. Looking to the long-term future it is interesting to question whether the same levels of mobility of goods and people will be maintained. That answer depends on whether industrial production continues to be organized on a global scale to take advantage of specialized skills in different places, whether individuals maintain contacts with friends and family across the globe and across a nation as well as within regions of the metropolitan area, and whether employees will still explore a variety of residential locations while valuing connections to a choice of workplaces. So long as some of those aspects remain influential in our economy and community, the networks that surround the polycentric structure described above will generate a strong underlying demand for transport spread across large metropolitan regions. Meeting that demand will call for much more functionally and spatially coordinated approaches to transport planning than have occurred in the past.

3 The global environmental crisis of transport

Patrick Moriarty and Damon Honnery

Introduction

Road vehicular travel is now in its second century. From almost zero vehicles on the world's roads in 1910, total vehicle numbers – those with four or more wheels – had grown to 1,003 million in 2007, of which 823 million – including sports utility vehicles in the US – were passenger vehicles (Altschuler et al., 1984: 2; Organization of the Petroleum Exporting Countries (OPEC), 2010: 81–90). By 2030, OPEC expects total vehicles to number over 1,700 million, including 1,370 million cars. US transport researchers Daniel Sperling and Deborah Gordon (2008: 4) also foresee similar vehicle numbers in 2030.

There is little doubt that such figures reflect the intentions of both the governments and citizens of the world, as discussed in Chapter 1. Populous countries like China and India, which presently have very low car ownership, wish to move to high-mobility societies, based on high ownership levels of locally manufactured vehicles. Given that most of the world's population still lives in countries with fewer than 25 cars per thousand population – compared with more than 400 per thousand population in OECD countries – there appears to be plenty of room for growth (Document 3.1 on this book's Companion Website expands on this theme).

At present, global vehicular travel is mainly constrained by incomes and, to a far lesser degree, by lack of road and parking space in some densely populated cities. In this chapter other possible constraints on future road travel are examined. Two are then selected – global oil depletion and global climate change from greenhouse emissions – as having crucial impacts on the prospects for future car travel. The various technical solutions that are proposed to overcome these constraints are then examined – namely improving the efficiency of transport vehicles and shifting to carbon neutral fuels – and we find that they will only be of marginal help. Given the limited value of such technological fixes, we then argue that a low-transport future is needed, and sketch out how this might occur. Finally this chapter discusses what urban transport might look like in the decades to come.

The global challenges facing road transport

Although this section will largely focus on the severe problems that oil depletion and global climate change present to transport worldwide, it will also examine three other challenges: road vehicle traffic casualties; air pollution from road vehicles; and the local environmental effects of oil extraction. Together, these are already widely acknowledged to be killing more than a million persons annually worldwide, with tens of millions more either seriously injured or made ill, and with disruption of societies and environments. Unlike the oil depletion and global warming issues, there is no disagreement about the basic facts. We have simply become used to this annual sacrifice for the doubtful benefits of high mobility. Even more importantly, only a small share of these high costs are being borne by the wealthy OECD countries.

Traffic casualties

Globally, road traffic fatalities in 2002 numbered over 1.2 million and a further estimated 50 million were injured (Ameratunga et al., 2006: 1533). The fatality rate per road vehicle varies inversely with vehicle ownership, so that, paradoxically, most fatalities are in countries with low car ownership, with Africa the continent with the highest death rate per vehicle. The explanation of this paradox is that low car-ownership countries tend to have very high numbers of unprotected road users – pedestrians and cyclists, and, especially in Asia, motorcyclists.

Consider a simple example in which all travel along roads is done either on foot or in cars. Assume that only 5 per cent of travel is presently by car, but that this doubles to 10 per cent. The number of pedestrians travelling will be reduced only marginally, but they will all now be exposed, *ceteris paribus*, to twice the traffic injury risk. Conversely, when most travel is by car or other four-wheeled vehicles, only a small percentage of travellers use the unprotected modes. The end result is that a high car-ownership country like the US, with around 25 per cent of the world's road vehicles, has under 4 per cent of global traffic fatalities (OPEC, 2010: 83; US Census Bureau, 2010: 844). Further, in nearly all OECD countries, the annual number of road fatalities has been falling in recent decades.

What will happen if global car ownership continues to rise, as is usually predicted? Given that vehicle ownership cannot be expected to rise much more in the high-mobility OECD countries – in fact, registered car numbers in the US fell by four million in 2009 over the previous year (Moriarty, 2010: 1) – the rise will come largely from higher ownership in presently low-mobility countries such as China and India. With the higher traffic fatality rate in low car ownership countries discussed above, global fatalities can be expected to rise disproportionately. Kopits and Cropper (2005: 176) studied the relationship between traffic fatalities and economic growth forecasts. They projected that global traffic fatalities would rise about 66 per cent between 2000 and 2020, with a 30 per cent decrease in wealthy OECD countries being more than offset by an 83 per cent increase in poorer countries. Growth in poorer country fatalities was projected to continue until their per capita GDP reached US$8,600 in 1985 values.

Air pollution

Air pollution is a major cause of mortality and sickness in cities. Traffic emissions are an important component of the urban air pollution problem, particularly in Asia, for two reasons. First, vehicle emission standards are often less stringent than they are in the industrialized countries of the OECD, and enforcement of existing standards may be less effective. Second, Asian cities have far greater population densities than Australian or North American cities (Newman and Kenworthy, 1999: 94), so many more people are affected by a given pollutant concentration.

This combination of a high density of both polluting vehicles and people in large cities can be a deadly one. In Asia, it was estimated in 2006 that outdoor air pollution was the cause of half a million deaths annually. Sixteen of the 20 most polluted cities in the world are in China, with vehicle emissions increasingly the main source of urban air pollution. Pollutants come not only from tailpipe nitrogen oxides, hydrocarbons and lead, but also from small particulate matter – suspended road dust, and particles from brake linings and tyres (Woodcock et al., 2007: 1081). Even in the UK, the urban residents at greatest risk from air pollution live in areas with the lowest car ownership. Although tailpipe emissions are anticipated to decline, increased hours per day in traffic, and the prevalence of cold start short trips – when emission control devices are least effective – will mean that adverse health effects will reduce much more slowly.

Environmental costs of oil extraction

As the Deep Horizon oil spill in the Gulf of Mexico very visibly demonstrated, there are heavy environmental costs in producing as well as combusting transport fuels. Because of progressive exhaustion of onshore oil reserves, attention is increasingly focused on the less accessible resources in deep offshore waters. In the US, as in other circum-polar nations, there is renewed interest in exploration of the Arctic Ocean. In the entire Arctic Ocean, reserves are estimated at 80 billion barrels of oil (Short and Murray, 2011: 162). For comparison, annual global production of oil in 2010 was 30 billion barrels of oil (BP, 2011: 8).

Short and Murray are blunt about the possible risks: 'A large spill in the Arctic could not be contained or mitigated, and we should stop pretending otherwise.' The nearest US Coastguard Station is 1500 kilometres away, and fog and snowstorms could hold up rescue work for weeks at a time. They further point out that:

> each major marine oil production area in the United States has seen at least one catastrophic spill: the 1969 blowout of a drilling rig off the coast of Santa Barbara, California; the 1989 *Exxon Valdez* tanker spill in Alaska; and the 2010 Deepwater Horizon blowout in the Gulf of Mexico.
>
> Ibid.

Exploration and production on land can also carry heavy ecological and social costs, particularly in countries with weak or poorly enforced environmental protection laws. In the head-waters of the Amazon 'spills of oil, drilling mud and other toxic components associated with the exploration and exploitation of oil and gas have polluted large areas of rainforests, including streams and rivers' (Jernelov, 2010: 360). As in the heavily polluted Niger delta in Nigeria, the local population receives few benefits from oil production, but must bear most of the costs of oil pollution (Moriarty and Honnery, 2008a: 869). Particularly in the Niger delta, the oil pipelines and other installations are often sabotaged, which along with illegal tapping of the pipes to siphon off products, further worsens pollution of the local creeks and land. The recently released report of the UN Environmental Programme reported, for example, that wells used by at least 10 Ogoni communities – the local people in the delta – had unsafe levels of hydrocarbons, and that the clean-up could take 30 years (Malakoff, 2011: 89).

Oil depletion

Despite much talk about biomass-based transport fuels and plug-in hybrids, road transport is still almost totally dependent on petroleum-based fuels. According to the International Energy Agency (IEA) more than 95 per cent of road transport energy is derived from petroleum, with nearly all the rest from natural gas and biofuels (IEA, 2011: 37).

The date at which 'peak oil' will occur is contentious, with all possible positions canvassed. Some argue that it has already occurred, others that it will occur in a few years, and still others that it is decades away, if it occurs at all. What is not disputed – although it is seldom discussed – is that on a global *per capita* basis, annual oil consumption peaked more than three decades ago in 1978 at about 0.7 tonnes, and from the early 1980s until the recent global financial crisis, was constant at close to 0.6 tonnes. Official bodies such as the US Energy Information Administration (EIA), the Intergovernmental Panel on Climate Change (IPCC), and OPEC are upbeat about oil availability, all assuming that the peak for oil production is decades off (for example, EIA, 2011: 231; OPEC, 2010: 63).

But when their forecasts for 2030 or later are expressed on a per capita basis, none of these organizations assumes that the per capita levels of the late 1970s will ever be reached again (Moriarty, 2010: 4). Nor do they expect any real rise over present levels, despite the low average level of oil consumption outside the OECD. It follows that if industrializing countries like India and China wish to increase per capita use, other regions will have to reduce their consumption, given the expected growth in world population. Global oil consumption is rapidly becoming a 'zero sum' game. There are signs that at least some of the major consuming countries already realize this (Klare, 2008: 1). Further, it is too easy to forget that future generations could also require oil and other scarce resources for essential uses (Gardiner, 2010: 58).

Actual new discoveries of oil in recent years are also much more in line with the more pessimistic view – represented by the Association for the Study of Peak Oil and Gas (ASPO, 2009: 2) analysis – than the optimistic view – represented by the US Geological Survey (USGS). Given the unreliability of world reserve estimates discussed above, the more consistent US data will be used as an illustration. In a widely discussed 2000 report, the USGS predicted that US discoveries of oil, condensate and natural gas liquids over the 30 years from 1 January 1996 would have a mean value of 159 billion barrels (Petzet, 2000: 79). Yet over the first 15 years of this 30-year period, discoveries totalled only about 38 billion barrels (BP, 2011), rather than nearly 80 billion barrels expected if, conservatively, annual discoveries were spread evenly over the entire 30-year period. In a recent paper, Miller (2011: 1573) gives a good summary:

> We agree that the world is not running out of oil, but the rate at which it can be extracted is falling. Oil will always be obtainable at a price, but that price may become unaffordable. Debates about whether the peak will be driven by inadequate supply, or by declining demand, are only different ways of stating the role of an affordable price.

Global climate change

Analysis of continuous cores painstakingly extracted from the Antarctic ice cap showed that the concentration of CO_2 in the Earth's atmosphere has varied from a minimum of 180 ppm (parts per million by volume) to around 300 ppm over the past 800,000 years. The lower values corresponded with glacial periods, and the high values with interglacials; calculated temperatures moved closely in step with CO_2 concentrations (Solomon et al., 2007: 24). By 2011, CO_2 levels were over 390 ppm, and are now rising at about 2 ppm annually.

As far as concentrations of CO_2 – and other greenhouse gases such as methane and nitrous oxide – are concerned, we are moving into unchartered territory. The concern is that Earth's climate will do likewise. The climate conference in Copenhagen in late 2009 (UN, 2009: 1) endorsed the European Union target of restricting the average global temperature rise since the start of the industrial era to 2 degrees Celsius (2 °C) in order to avoid 'dangerous anthropogenic climatic change'. The question of what is actually 'dangerous' and to whom or what is itself a controversial matter. A 2 °C rise might be survivable by the majority of humanity, though the negative impacts will be unevenly distributed around the globe. But such a rise in temperature, if it occurs over the short time horizon of a hundred years or less, is likely to do great damage to the planet's biodiversity. Many species will not survive local habitat change and will become extinct. Whole ecosystems such as coral reefs and rainforests will be seriously threatened (Le Page, 2011a: 43; Warren et al., 2011: 150–54).

This cap on allowable temperature increase gives us very little leeway, since temperatures have already risen an estimated 0.76 °C, and we are committed to a further 0.6 °C rise because of the thermal inertia of the oceans. So, even if we ceased all emissions today, the Earth average temperature will eventually rise to 1.3–1.4 °C above the pre-industrial level. And given that more than 80 per cent of the world's energy needs are still met with fossil fuels, we also need to consider institutional inertia in emissions. We evidently have very little room to manoeuvre if we take the 2 °C limit seriously.

Of course, in a system as complex as the climate system, with its multiple feedbacks acting in different time frames, there are great uncertainties, both about the level of greenhouse gas concentrations that we can allow while keeping temperatures below the limit, and whether the limit itself is set too low. We can expect to see the limit steadily raised as we approach, and pass, the 2 °C limit. Nevertheless, if we assume this limit, then modelling shows that the world would have to cut CO_2 emissions in the year 2050 by as much as 85 per cent of their year 2000 value. This ambitious reduction target would mean that global annual fossil fuel use would have to be limited to about 50 EJ (EJ = exajoule = 10^{18} joules), rather than the 437 EJ we used in 2010 (BP, 2011: 41; Moriarty and Honnery, 2009: 34).

In 2009, Australian *per capita* emissions of CO_2 from fossil fuel combustion and cement manufacture were 17.9 tonnes, or 4.2 times the world average (IEA, 2011: 48). If emissions were to be reduced to 15 per cent of year 2000 levels, and equal for all the projected 9.15 billion humans in 2050, allowable emissions would only be 0.377 tonnes per capita (BP, 2011: workbook). Australian 2009 CO_2 emissions would need to be cut by a factor of 47. Cuts in US emissions would be similar, but less for other OECD countries, typically 25–30 fold. Put simply, our emissions would need to be cut to near zero. At 0.377 tonnes per capita, even industrializing countries such as China, India, and Indonesia would see large reductions in CO_2 emissions (IEA, 2011: 48).

One prominent climate researcher, James Hansen, argues strongly that even a 2 °C limit, though very difficult to achieve, is itself too high. In a recent paper (Hansen et al., 2011: 5), he claimed that the paleoclimatic data point to a limit of only 1 °C above pre-industrial if irreversible destabilization of the polar ice-caps is to be avoided. Or as Michael Le Page (2011b: 41) put it: 'In comparable interglacials in the past half million years, when temperatures were less than 1 °C warmer than they are now, sea level was around 5 metres higher.' As we are already committed to exceeding this temperature rise, it follows that even the existing CO_2 level of over 390 ppm is too high, and that it should be *reduced* to 350 ppm. Not only would annual emissions have to drop from their 2010 level of over 33 billion tonnes from fossil fuels alone (BP, 2011: workbook), but some past emissions would also have to be removed from the atmosphere. Of course, in our present political climate, such cuts are wildly infeasible. It is all too easy implicitly to think that because an action is politically infeasible, the problem will disappear. Unfortunately, the natural world doesn't work that way.

Regardless of the CO_2 levels that are agreed, transport will not be exempt from these emission cuts. On the contrary, it is likely that urban passenger transport will have to make larger reductions than the economy overall. In the transition to far lower oil use and emissions, freight transport can make claims for priority in delivering food to cities and raw materials and intermediate products to industry. Without freight, industrial economies would soon collapse. Cities in the West have only three days' food supply, and so must be continuously resupplied. Even non-urban travel could claim priority over urban travel, given that alternative modes are more difficult to organize, especially for remote settlements. Accordingly, the remainder of this chapter will examine the implications for urban passenger transport, mainly in an Australian context.

Proposed solutions: technological fixes

Possible technological fixes include improving the fuel efficiency of conventional road vehicles, both passenger and freight, greatly expanding the use of alternatives to petroleum fuels, and the development of new propulsion systems. Although none of these helps reduce traffic casualties, they do promise simultaneously to tackle oil depletion and air pollution, both local and global – from CO_2 and other greenhouse gases.

Improving vehicular energy efficiency

The energy efficiency of a vehicle is usually measured by comparing the vehicle-kilometres of travel for a given fuel input, as in the familiar 'miles per gallon', or in today's units, kilometres per litre of fuel. The best way to think about energy efficiency is to look at what the energy in the vehicle's fuel tank is actually used for. Some is used for lighting, air-conditioning, powering devices such as windscreen wipers, vehicular entertainment and information systems, and for power steering, braking, etc. But most is used to overcome a vehicle's resistance to forward motion. This resistance, or *road load*, has three components. The first, rolling resistance, is the frictional resistance between tyres and the road. This resistance can be reduced by better tyre design and higher tyre pressures – but higher pressures give a less comfortable ride.

The second component is air resistance, which is important at the higher speeds attained in highway and freeway driving. It can be reduced by better attention to aerodynamic design – and, of course, by reducing driving speeds. The third component is the most important for inner city driving: overcoming vehicle inertia, that is, getting the vehicle to a higher speed from either a stationary or a lower speed state. Like rolling resistance, inertial resistance varies directly with vehicle mass, so reducing the loaded mass of the vehicle can reduce inertial resistance. Smaller vehicles are thus more fuel efficient than large ones. Cutting down the amount of stop-start driving will also reduce the road load.

All the energy supplied to the wheels of a road vehicle – whether car, truck, or bus – is used to overcome this road load, so reducing it will increase vehicle fuel efficiency. But another approach is to improve the engine and drive-train efficiency, that is, increasing the share of the input fuel energy in the tank that results in propulsive energy at the wheels. Over the past century, engine efficiency has been continuously improved, but has resulted in little overall improvement in kilometres per litre. The reason is that the energy demands on the vehicle have also risen, negating the engine efficiency progress. The road load has risen, mainly because of the shift to larger, heavier vehicles. So has the energy needed to run the auxiliary equipment: air conditioning, power steering and the like.

The overall result is that there has been little improvement in fuel efficiency worldwide. Woodcock et al. (2007: 1085) reported that present-day Ford cars in the US were no more efficient than the model-T Ford. For the Australian car fleet, our own analysis showed that over the two decades 1985 to 2005 there was no gain in fuel efficiency (Moriarty and Honnery, 2007: 5).

Can we do better? As could be inferred from our discussion, maximum efficiency gains will come about from combining both drive-train efficiency gains with reductions in road load. Inertial resistance dominates road load under stop-start urban driving conditions. A shift to electric drive vehicles, as in the hybrid vehicles presently being sold – such as the Toyota Prius – promises to reduce road load by regenerative braking. Here, instead of the kinetic energy of the moving vehicle being converted to heat and dissipated by the brakes, it is converted into electrical energy and stored in batteries for re-use later (Moriarty and Honnery, 2008b: 1720). Also, because hybrids have both an electric drive motor and an internal combustion (IC) engine, it is possible to use a smaller-sized – and so lighter – IC engine, and run it under more constant load conditions. The result is that hybrids can be much more efficient than conventional IC vehicles under urban driving conditions.

If both all feasible road load reductions and hybrid drive-trains were implemented on future vehicles, a Swedish review (Akerman and Hojer, 2006: 1946) estimated that a four-fold rise in fleet fuel efficiency for urban driving might be possible by 2050, and a threefold rise for highway driving. By 2030, it might, at best, be possible to double car fleet fuel efficiency (Moriarty and Honnery, 2008b: 1722). Compared with the dismal history of fuel efficiency gains, a three- or four-fold rise seems very optimistic. But when compared with what is needed for climate change mitigation by 2050, even improving efficiency by a factor of four is nowhere near enough.

Two additional factors must be considered. Since presumably the aim is to move people, not vehicles, we should really consider passenger-kilometres per litre of fuel, not vehicle-kilometres per litre. Urban car occupancy rates are typically about 1.5 or less today, so there is much scope for improvement in typical five-seater cars. But in Australia and other high-mobility countries, occupancy rates are falling, and we will be doing well if we can prevent further falls in the face of declining household sizes and rising car ownership (Moriarty and Honnery, 2012: 1963). The other factor concerns the energy required to get a litre of fuel to the fuel tank. As conventional oil is depleted, the world will increasingly have to rely on unconventional oil in future. But deepwater oil, and oil from tar sands or shale oil deposits have far higher energy needs and greenhouse gas (GHG) emissions per unit of fuel delivered to the fuel tank. Like falling occupancy rates, these higher energy needs for fuel production will counteract any hard-won improvements in vehicle fuel efficiency.

Shifting to alternative fuels/power systems

If we cannot expect much in the way of reductions in either oil use or GHG emissions from improving vehicular fuel efficiency, what about changing the fuel? Not only will oil be expensive and in more restricted supply in future, but, as we have just seen, its GHG emissions per litre of delivered fuel, already a problem, will get worse. Moving to clean – or 'green' – fuels is an obvious answer. Such an approach promises to solve the oil depletion problem, and both local and global pollution.

The use of non-petroleum fuel for transport is growing, with the leading countries Brazil and the US. In Brazil ethanol is made from sugar cane, and in the US from corn – and sorghum. In the EU, although some ethanol is made from grain, the preferred alternative fuel is biodiesel, made from edible oil seeds such as rapeseed. Global production is minor compared with ethanol. Worldwide, liquid biofuels production in 2010 was the equivalent of 69.1 GL (gigalitres) of oil, or about 2.8 per cent of all refined transport fuel including non-road transport (BP, 2011: 39; IEA, 2011: 39). Since all transport biofuels are presently based on food crops, it is recognized by all that further expansion of biofuels will soon need a move to cellulosic ethanol.

Liquid transport fuels based on edible crops not only raise ethical questions, but are only of limited help in reducing transport GHG emissions. Although US corn ethanol is often reported as giving roughly a 20 per cent – not 100 per cent – saving in carbon emissions, two factors not usually considered actually negate any climate change advantage, even if it does replace imported oil. First, corn production in the US requires nitrogenous fertilizer, and emissions of nitrous oxide – an effective and long-lived greenhouse gas – from the soil alone negate any carbon emission savings (Crutzen et al., 2008: 389).

Second, as Searchinger et al. (2008: 1238) have pointed out, there are flow-on effects in food-importing countries from using US corn for fuel rather than food. When the global consequences are considered, the authors again found that corn ethanol produced higher GHG emissions than petrol. Using cellulosic ethanol from switchgrass grown on corn land was also argued to produce higher emissions. Because of the system considerations discussed here, it is likely that cellulosic ethanol – which is not yet in commercial production anywhere – from biomass plantations, or even from agricultural and forestry wastes, will produce little or no net GHG emission savings, as Felix and Tilley (2009: 410) also argue.

As discussed above, electric drive is an efficient means of propulsion because it allows for regenerative braking. A further advantage from a climate change mitigation viewpoint is that the propulsive electrical energy needed can be derived from a variety of sources, including renewable energy (RE). In fact the primary output for many RE sources – wind, solar photovoltaic, hydro – is electricity. At present, the only road vehicles with electric drive are hybrid electric vehicles, but the electric power produced is ultimately derived from the petrol tank. Future vehicles, however, could be powered by mains electricity.

Several options are possible. The most popular one at present is plug-in hybrids, which would have two energy sources: a conventional petrol tank, and a battery pack charged overnight from mains electricity. It is also possible to have a fully electric vehicle, without a petrol IC engine, but the battery requirements would then be excessive – and very expensive. For plug-in hybrids, the idea is to size the battery so that the vehicle can go, say, 40 kilometres without recharging. For days with greater travel needs, the motorist could then use the petrol engine.

Developing a battery with sufficient energy density – energy stored per unit mass of battery – has been the main difficulty for both fully electric and even plug-in electric vehicles. The batteries must not only have high energy and power density, but must be made from non-toxic materials that are available in adequate supply, be safe to use, be able to withstand thousands of charge/discharge cycles, have low energy conversion losses, and be not too expensive. No battery will ever come near the energy density of petroleum fuels, but batteries based on the lightest metal, lithium, are being promoted as the answer for plug-in hybrids. As discussed in Moriarty (2010: 9), lithium-based batteries are not only very expensive, adding as much as US$18,000 to the cost of a vehicle, but, at least according to OPEC, also could not support more than a small fraction of the large projected vehicle fleet discussed above in the first section because of the limited lithium resource base. Others dispute this lack of lithium reserves. The question may not even matter, since the high cost and safety problems of lithium-based batteries means that research on other battery types continues.

A more radical approach would use hydrogen as a fuel in hydrogen fuel cell vehicles (HFCVs). The hydrogen, if derived from electrolysis of water, could eventually be based on electricity from renewable energy (RE), but at present hydrogen is produced from natural gas. A large number of difficulties are yet to be overcome for HFCVs to be a viable option, including the high cost but limited life of the fuel cell pack, and limited availability of the platinum used for catalysing the reactions at room temperature. But just as serious is the difficulty of storing hydrogen onboard in cars, although this is far less of a problem for large vehicles, including trucks, buses and trains. Not only are there serious safety and other obstacles with storing hydrogen as either a highly compressed gas or a very low temperature liquid, but the energy costs of compression/liquefaction and transport to refuelling stations will be high (Bossel, 2006: 1829). Using hydrogen directly as a fuel in IC engine vehicles would face the same problems of hydrogen onboard storage and transport.

Although, like any alternative fuel, hydrogen would help overcome the oil supply problem, it would only be advantageous from a GHG viewpoint if made from RE electricity sources. Yet a marked shift to RE for electricity will not happen anytime soon. At present, the share of fossil fuel in global – and Australian – electricity generation is still rising slowly, and official forecasts are that this situation will change little over the next two to three decades (EIA, 2011: 254). And until such time as carbon-neutral energy accounts for almost 100 per cent of electricity generated in a grid – as is true only for some small countries like Norway and Iceland – GHG production in the economy overall will always be less if such electricity is used *directly* in end-use applications. So applications such as for traction power for electric public transport, or for electronic equipment should have priority over battery or hydrogen-powered vehicles, where a large fraction of the generated carbon-neutral electricity will be lost in conversion, for example to hydrogen.

To summarize, neither vehicle energy efficiency nor alternative fuels looks likely to offer dramatic reductions in either oil use or GHG emissions in particular. Even on a passenger-kilometre basis, it will be difficult to achieve much in the way of reductions because of declining vehicle occupancy rates, declining shares of carbon-neutral energy for electricity generation, and increased reliance on non-conventional oil. Population increase will further offset reductions on an absolute basis. For Australia, at best a threefold reduction in passenger car GHG emissions per kilometre might be possible by 2050, but, as we have seen, a 50-fold reduction might be needed. As Rees (2009: 300) has argued, putting our faith in technological fixes for transport problems promises to make matters worse by deflecting attention from the changes really needed.

Proposed solutions: social changes

Most discussions of the need to reduce either transport oil use or GHG emissions assume that only technology-based solutions are possible. On the other hand, reductions based on social changes are not rated highly (see for example, the IPCC in Kahn Ribeiro et al., 2007: 360). Infinite possibilities for technological change are assumed, but very limited possibilities for social change – except for those needed to adjust to the new technology. Although such a position closely matches corporate interests, the preceding section showed that technological fixes will not give us anywhere near the needed cuts in either oil or GHG emissions.

The changes needed are of two kinds. First, we will need to reduce our levels of vehicular travel, as measured by passenger-kilometres per capita. Second, we will need to move nearly all the remaining vehicular travel to public transport. Many people seem to believe that shifting to public transport will be enough, because it is believed to be vastly more efficient than car travel. But on a primary energy per seat-kilometre basis, the public transport modes are typically only about two to three times as efficient as car travel (Moriarty and Honnery, 2011: 192).

However, several other factors need to be considered, and they all act to shift the balance towards public transport. The occupancy rate – the proportion of seats filled – for cars in OECD countries is today typically about 30 per cent, or 1.5 persons per 5-seater car. Public transport occupancy rates in Australia, though rising, may not be much higher than this at present, but they can potentially be much higher – as in Tokyo. There is, of course, a downside to all this: the transport planner's high occupancy rate is the commuter's grossly overcrowded peak hour train.

Another undoubted benefit of electric public transport is its independence from oil. It is likely that urgent action will be needed to curb oil imports before serious emission reductions are imposed to mitigate global warming.

The most important – and least appreciated – factor is that total vehicular passenger travel levels are not independent of the dominant travel mode in a city. Several lines of evidence support this conclusion. First, no city in the world with public transport as the dominant mode has ever approached 5000 passenger-kilometres per capita of public transport, except Tokyo, with 5500 in 1990 (Newman and Kenworthy, 1989: 342, 1999: 84). One obvious objection is that public transport-dominated cities have either been low-income cities – for example, Australian cities in the 1940s and 1950s – or very densely populated, or both, as was the case for Japanese and other Asian cities up to the 1960s. To answer this objection, we first need to consider travel times for different modes.

Public transport modes are inevitably slower on a door-to-door basis than car travel, except, perhaps, for trips to or near the city centre. Reasons for this include the need for public transport to stop frequently to drop off and pick up passengers, the time delay for any trip interconnections, and the circuity of many non-central public transport trips.

Passengers are conventionally regarded as considering two costs of travel in their decision making: the monetary costs of travel, and the time costs of travel (Moriarty, 2002a: 3). Car travel may be more expensive on a fully monetary costed basis than public transport – although the *perceived* costs for any given trip may be less – but the time costs will be far higher for most trips if undertaken by public transport, at least as it presently exists in most cities.

As urban populations become wealthier, more and more of them will become time-constrained rather than money-constrained, and will prefer the faster mode. So even given relatively cheap travel, if only slow travel modes are available, they will travel much less. This time factor, rather than income, largely explains why travel levels in cities with high public transport use will often be lower than in car-oriented cities. Paradoxically, a further consequence may be that *overall* travel times are less in public transport-oriented cities. Such appears to have been the case for Melbourne as it shifted from public to private travel (Moriarty, 2002a: 5). To summarize, travel costs have both a money and time component; if either is reduced, people will travel more.

Third, although travel is usually seen as a derived demand, not all travel is undertaken solely for utilitarian purposes. For many trips, it is truly 'better to travel than to arrive'. Particularly for young males, and even for others, mastery of a car provides psychic – i.e. non-instrumental – benefits. Even for the commuting trip, many drivers see it as a time when they can be alone – 'quality time' in today's jargon. Kaufmann (2005: 50–51) showed from his European data that the way people value travel – by different modes – varies considerably. He identified four types of travel valuation behaviour, which he defined as follows:

- *Efficient*: people who want to get from trip origin to destination in the shortest possible time.
- *Sensorial*: people who value the experience of the travel for its own sake – for example, walking or cycling – related to experiencing the environment travelled through.
- *Pre-planning*: people who minutely determine travel in advance, in which travel is integrated with valued activities – working or relaxing on the train, music in the car.
- *Flexible*: people who react to opportunities that present themselves in the course of travel.

Fourth, not only trip times for any given trip, but also tripmaking patterns will vary with the dominant mode. At present in the car-oriented Australian cities, car travel calls all the shots. Since around 90 per cent of vehicular travel is by car, most destinations are now those than can be easily accessed by car, not alternative modes. Travellers using other modes have to fit in as best they can. What this means is that the inevitable difference in door-to-door trip times is exaggerated. If the slower public transport modes were dominant, there would be far more incentive to combine trips for different purposes. At present the speed of car travel allows activities to be *interwoven*; parents may chauffeur their children to Saturday morning sport (as discussed in Chapter 10), then go and do some shopping before returning to pick them up. Because of time constraints, such interweaving may simply not be possible for public transport – and for older children, chauffeuring would simply not be done.

An important question arises: Can the existing car-oriented cities of Australia and elsewhere be adapted for travel by alternative modes? Will this prove impossible without massive restructuring of the urban form? We do not think so. In Chapter 9 the means of such adaptation are proposed. A look at the changing urban structure of Australian cities supports our contention that adaptation is possible. At the time of Australia's first post-war census in 1947, the capital cities were highly centralized. Most jobs and retail sales were usually in the municipality containing the city's CBD, and the electric tram and rail systems were oriented towards the city centre. Up to 80 per cent of vehicular travel in capital cities was by public transport, mainly trains and trams (Moriarty and Honnery, 2005: 2). With both destinations and dominant travel modes oriented towards the city centre (as also noted in Chapter 2), the result was that most vehicular trips were similarly oriented. Average vehicular trip lengths were thus strongly influenced by the average resident distance from the CBD, which was naturally higher in bigger cities.

In contrast, today's cities have a much more even spatial balance of residences, workplaces, shopping and leisure centres, and other services. So although vehicular travel per capita in Melbourne, for example, is today about four times its year 1947 value, travel *needs*, as measured by the separation of residences from jobs, shops and other destinations, have actually reduced (Moriarty and Honnery, 2008a: 870). What this suggests is that although car-oriented cities have far greater travel levels, this is a result of the higher travel *convenience* and faster door-to-door speeds of car travel. But the type of urban form tomorrow's public transport systems will inherit are *potentially* low-travel cities. Further, the intermixing of land use functions makes walking and cycling possible for many trips.

A possible future for urban transport

We will now try to give some idea of what shape passenger travel in Australian cities will likely take in the coming decades. First, it must be recognized that any serious attempt to tackle global warming will require reductions in *all* sectors of the economy, not just transport. At first glance, oil depletion seems mainly a problem for the transport sector, given its heavy reliance on petroleum products. However, global oil depletion will have significant effects on the global economy more broadly (ASPO, 2009: 3). Change in all economic sectors means that the nature of work, the type of jobs people do, will also change.

Particularly if economies worldwide become more oriented towards satisfying the human needs of all the world's people (Moriarty and Honnery, 2011: 187), it is difficult to see much future in Australian cities for today's heavy emphasis on banking, finance, real estate, marketing, and the obsession with retail sales. This type of employment is concentrated in the centres of our large cities, and the loss of many of these jobs will mean a loss of public transport patronage from these job-holders. Other functions of the city centre will survive, and new types of employment can be expected to develop. The centre will remain the most accessible region of the city, and is the logical place for the location of major sporting, entertainment and education centres, city – or even State-wide – administration offices, specialist shops and services. Increased public transport travel to these destinations will offset much of the patronage loss just discussed.

It is also important to recognize that future urban transport systems will have to take the urban form largely as given; because buildings and urban infrastructure have a lifetime of many decades, there is not the time for extensive modifications. Many writers have seen urban densification as a key factor in making urban transport more ecologically sustainable. Even if such densification helped urban transport, and there are doubts about whether such changes are necessary, it might not help – and could even hinder – other components of urban sustainability. Future cities, for example, may need to be more self-sufficient in food – mainly vegetables and fruits, not cereals – water and even energy provision. The existing lower density settlement patterns of large Australian cities could even help here. What is optimal from a sustainable transport perspective may conflict with sustainability in other sectors; compromises may be necessary. Even so, it makes sense to have higher residential and workplace densities in the central area and around public transport nodes such as railway stations. It is even possible that the large metropolitan centres could lose some population to rural areas, given the growing demand for food globally. If this happens, it would further slow the pace of urban change.

The changes to employment, or, more generally, to a 'human needs economy' will not happen rapidly. Since we do not really know what the urban future will be, we will have to feel our way. Although large-scale change to the existing urban fabric is unlikely for the reasons just given, there should be enough time to implement many 'micro-changes' in the way it is used. For example, local shopping centres could be used much more intensively than at present, with accordingly less use of drive-in shopping centres. Many existing workplace buildings could be used for the new job types that will inevitably arise in ecologically sustainable cities.

Australia's cities of the future will likely be low-mobility cities, a requirement that follows from energy and environment constraints. Given these constraints, much more *localization* of activities is probable – working, socializing and entertainment, shopping and so on (Moriarty and Honnery, 2008a: 871). Non-motorized transport is ideally suited to these local trips. For longer trips, bus travel could be more important than today because, in the transition, rapid changes and new combinations of bus routes are possible because of the vast networks of urban roads – our inheritance from the automobile age.

Part 2

Global and local change and persistence

4 From automobility to sustainable transport

Nicholas Low

Introduction

The problem for future urban transport emerges very clearly. It is a global dilemma, and unless the global perspective is maintained the dilemma will not be resolved. On the one hand, the global economy depends more than ever before on mobility. Intra- and inter-regional merchandise trade even after the global financial crisis stood at US$12.178 trillion in 2009 (WTO, 2010, Table 14): 12 trillion dollars worth of stuff moving around the world. There is, we can say, *embodied* mobility in almost everything we consume and produce today. On the other hand, that mobility depends overwhelmingly on fuels whose combustion is radically changing the biosphere through global warming. And it is time also to reconsider the more immediate and devastating personal costs of automobility to human health and life and to the health and life of local ecologies. A plague called traffic cannot be regarded as just a private matter. It is a public health issue. A recent article in the English language newspaper *China Daily* pointed out that 'in lower income countries the number one killer of people between 15 and 29 is road crashes' (Deng Fei, 2011). A recent epidemiological study of London and Delhi described the huge amounts of money that could be saved if sustainable transport came to replace automobility as the dominant paradigm (Woodcock et al., 2009).

The problem is global not only in the trivial sense that it occurs all over the world, but because the economic and ecological systems that ultimately make mobility possible are globally interlocking. For instance, it is no use imagining that electric cars will be a sustainable transport solution if, once taken up by the world economy – and especially by the emerging superpowers of China and India – it is found that there are not enough available mineral resources to build batteries for the world fleet. Nor is an electric – or hydrogen – solution sustainable if the power to generate the electricity or hydrogen is made by fossil fuels. So the first step in re-visioning urban transport must be to understand the fundamental problem as we have described it in the foregoing chapters of this book. The next step is to create an embracing vision of sustainable transport for the future that offers the hope of addressing the problem and resolving the dilemma. Before considering the dimensions of such a vision, however, we will review the vision we must leave behind, the vision which is still driving urban transport today along what we can now see is a road to perdition: automobility. It is a compelling vision, and it is worth considering not only what it stands for but the socio-technical process by which it was promoted and finally realized.

The cyclone of automobility

The vision of individual automobility is a phenomenon of the twentieth century. It offered freedom to move in unprecedented comfort, at unheard of speed, with an extraordinary level of personal control, an escape from the drudgery and danger of city life and the promise of reconnection with an idyll of nature. Moreover, after the mass production of cars was created by Ford and others, the vision was open to all to realize. The history of automobility is well documented in the USA, Europe and Australia, though different aspects of this history appear in different works. Sachs (1992), for instance, gives a particularly well-documented account of the rise of desire for independence, style and speed in Germany with less emphasis on the growth of the car industry. In the USA, Flink (1988) describes the rise of the automobile industry in great detail, but we look to McCarthy (2007) to tell us more about the consumption side: 'cars, consumers and the environment'. Davison with Yelland (2004) in Australia describe the political history of cars and roads. Dudley and Richardson (2000) examine road – and railway – policy in Britain from a political science perspective, and there are many other works touching on different aspects of the development of automobility (Wachs and Crawford, 1992; Sheller and Urry, 2000; Mosey, 2000; Vigar, 2002; Garrison and Levinson, 2006; Packer, 2008; Conley and McLaren, 2009). The historical record is therefore something like an incomplete jigsaw puzzle. The pieces for Asia and the new dynamic economies – for example, Brazil – are unfortunately missing, though there are some hints of significant local difference in the process, in Simon (1996), Dimitriou (1992, 2011), Karan and Stapleton (1997), Wu et al. (2007), and Imran (2010). But, however incomplete the picture, a pattern can be discerned which seems to apply at least in the industrialized world wherever automobility has evolved: a continuously self-reinforcing process in which public visions, private aspirations, technologies, industries, government regulation and infrastructure construction feed off one another, becoming powerful enough to overcome all countervailing forces to generate a structured path of human collective behaviour, something like a socio-technical cyclone.

How does such a cyclone start? No one can really know, but the conditions for its formation were certainly present for more than a century before the twentieth. Rifkin (2009: 335) tells us:

> As the tempo of industrialization picked up speed, urbanization accelerated and national governments consolidated their power, the 'rationalization' of government, market and social relations continued to keep

> pace. The period between 1790 and 1850 saw the increasing atomization of the individual in an ever more rationalized and integrated society.

Mobility and speed of travel were central to the integration of society composed of individual identities seeking pleasure and avoiding pain. The idea of speed was first realized with the steam locomotive, but with the invention of the automobile, speed could be appropriated individually by the rich. McCarthy (2007) begins his book on 'auto-mania' with a vivid description of the rapid adoption of the car for fun and sport by young American millionaires who 'owned and raced yachts, bred and ran thoroughbred, steeplechase, and harness racing horses, and played competitive polo, golf and tennis against one another. ... Speed had nothing to do with getting somewhere sooner. Driving was simply fun'. Less fortunate Americans, he writes, could only 'gape in awe, envy and exasperation' (ibid.: 1). Likewise in Germany, as Sachs (1992: Ch 1) reports, automobiles provided 'pleasures for the wealthy'. Driving was an adventure, and the driver a sportsman 'taking up the reins of technology' (ibid.: 6). Interestingly the advertised imagery of cars, tapping the wellsprings of human desire, has changed little since the early days: geographical escape; social status; individual freedom; gender and sexuality – masculine: with cars sometimes portrayed as extensions of masculinity and sometimes as love objects; and power, speed and excitement (Conley, 2009: 39–40).

So the first cycle turned from the invention of a technology – automobiles, whether powered by steam, electricity or petrol (gasoline) – to its early adoption by the rich, and the creation of widespread envy and aspiration. As the reality and potential of cars became widely understood, there was strong resistance. A future President of the USA, Woodrow Wilson – then President of Princeton University – foresaw with deep consternation the outcome of automobility: 'I think, of all the menaces of today, the worst is the reckless driving in automobiles ... In this the rights of people are set at nought'. Wilson asks how a person would react to the death of his child from being run over: 'I don't blame him if he gets a gun' (27 February 1906 cited by McCarthy, 2007: 12). McCarthy states:

> Wilson's remarks seem shocking, especially coming from a future U.S. president, but in 1906 they were typical of anti-automobile statements made by many elites, and compelling evidence not of a cavalier attitude on the part of Americans towards vigilante violence, but of the emotional intensity of the reaction to the automobile.
>
> Ibid.: 12

Similarly, in Germany, Sachs (1992: 13) reports the antagonism in the early twentieth century to the potential for cars to dominate the streets, especially of rural towns, which 'were after all inhabited by pedestrians, horse drawn conveyances of various types, children at play, and all kinds of fowl'.

In 1900 cars were banned from the streets of the rural Swiss canton of Graubünden.

Almost simultaneously, however, the aspirations of the many were turning people into consumers of automobiles with real interests just as the craft of automobile manufacture was becoming an industry with production interests. People wanted not just cars, for the reasons cited above, but also everything necessary for car travel: better roads, more available and cheaper fuel, parking space, accessories. As Flink (1988: 28) citing Interrante (1983: 90) observes, the consumption of cars 'satisfies a real need for transportation – a need as basic as food, clothing and shelter'. Aspirations of excitement gradually gave way to 'utilitarian advantages of economy and efficiency over the horse', which was, after all, the prime mover of the medieval transport revolution (Flink, 1988: 28). The British historian Tony Judt, who spent much of his working life in the USA, points out that the 'car culture' came to Europe in the 1950s, but, he writes, his own post-war generation 'grew up with cars and saw nothing distinctively appealing or romantic about them' (Judt, 2011: 42).

Car manufacturers wanted to make profits by selling cars. The petrol engine was first perfected in France and Germany giving French and German industries a brief advantage, but the principles of internal combustion were quickly taken up in Britain and the United States as cars and engines were imported. By 1904 American car production had overtaken that of France: 'France lost technological leadership as well to the United States with the appearance of moderately priced, light American runabouts in 1906–1908' (Flink, 1988: 25). Car manufacturers rapidly grew to be major employers. Ford wasn't the only manufacturer to see that higher wages would produce automobile consumers. The growth of consumption and production interests was accompanied by promotional activities such as car races, and, most importantly, the growth of a specialized press 'crusading' for mass adoption of the automobile (ibid.: 29). Flink noted the part played by the motoring clubs on both sides of the Atlantic in 'sponsoring tours and tests, lobbying for legislation favorable to motorists, and propagandizing the automobilists' cause' (p. 27) – and the 'close cooperation' between the press and the automobile industry (p. 30).

There is no need for conspiracy theories to explain the pressure put upon governments to support such a rapidly growing and massively employing industry, support mainly for building new and better roads for the cars to use: the 'good roads movement'. Flink reminds us that farming and rural interests in the USA were among those lobbying for better roads. At the Second Annual Good Roads Convention in 1909 at Cleveland, Ohio, the dominating influences 'were the American Automobile Association, representing the autoists and the cities, and the National Grange representing the farmers. Cooperating with these organizations were the American Road Makers' Association, the National Association of Automobile Manufacturers, and the American Motor Car Manufacturers' Association' (Flink, 1988: 170). But road building lagged demand. Not until 1930 was the primitive American road network, mostly unsurfaced and poorly signposted in 1904, transformed into 'an interconnected system of concrete highways' (ibid.).

Similar poor quality rural roads – in terms of the needs of the car – were normal all over Europe up to the end of the 1920s. Roads were, 'planned according to small scales, laid out for slow speeds, twisting along creeks and over hills, and emptying directly on to a market square, these streets were of use to bicyclists and horse carts but not fit for the space-mastering power of the automobile' (Sachs, 1992: 45). Sachs tells the story of the pressure of automobilism in Germany similar to that of America: 'the people's car and the freeway – the two pillars of a society on wheels' were to be realized in the 1930s, 'but not under the sign of democracy' (ibid: 46). The idea of a road just for automobiles linking cities across the country had been promoted by the construction industry in Germany in the 1920s, but Hitler's Reich led the way on motorway construction after 1933 with a view to stimulating the economy, creating jobs and unifying the nation.

After the War, America began the construction of a massive programme of interstate highways under the Interstate Highway Act of 1956, financed from the gasoline tax and other taxes on cars. Expenditure on roads between 1947 and 1970 was estimated to take up 75 per cent of all government expenditure on transportation (Flink, 1988: 176). Road building programmes were initiated in Britain, France and Italy as part of post-war reconstruction, and road building required expertise. Dudley and Richardson (2000: 86) note that the Minister of Transport, Boyd-Carpenter, 'sent delegations of engineers to Germany, the United States, France and Italy and so provided leadership to what became a key epistemic community, whose expertise would be required if the roads construction programme was to be accelerated'.[1]

In the 1920s only 2 or 3 per cent of people in the State of Victoria (Australia) held a driving licence. By 1950 most people in Australian cities still travelled around on foot, by bicycle or on public transport. It took a strong, sustained campaign by far-sighted people to convince governments that travel by car was the way of the future. Australian historians Davison and Yelland write about 'how the car won our hearts and conquered our cities' (Davison with Yelland, 2004). 'Motorists', they acknowledge, 'were a noisy lobby group, never reluctant to blow their own horns. Their opponents were a more diverse and defensive group, steeped in traditional attitudes and institutional loyalties, conscious of their declining patronage and often forced on to the back foot by the aggressive propaganda of motorists' (ibid.: 125). Australian town planners and road engineers visited the United States and returned with visions of freeways sweeping through the countryside. There was a strong aesthetic dimension: the sublime of engineering. Davison and Yelland (2004: 171) report the words of the Chairman of the Country Roads Board of Victoria: 'I really believe the freeway is a work of art ... they are beautiful things, not evil things. They are things which enhance the beauty of the countryside not desecrate it'. But this was said in

1974, when building motorways through cities was under challenge from community groups who did indeed see the motorway as evil and the instrument of desecration.

By the time urban freeways were being built – and challenged – in the 1960s the problem of providing infrastructure for cars had changed radically. It was no longer one of bad roads but of traffic congestion on mostly very good roads. Good roads, cheaper cars and higher incomes enormously increased the volume of traffic, and with it the number of deaths and injuries from vehicle crashes. Cars, which were at first thought of as fun vehicles for speed and escape to the countryside, were replacing public transport for the journey to work in cities. New communities of experts grew up around the twin problems of road safety and the analysis of congestion.

When Woodrow Wilson warned of the future risks of a car-borne society he formulated the problem in terms of road safety and irresponsible driving which points to a solution: making roads safer to drive on, educating and regulating drivers, and making cars safer to drive *in*. This formulation does not challenge the idea of automobility itself, linked as it is with individual liberty. Road safety campaigns have been carried out in some form everywhere in the developed world alongside the growth of automobility. Mostly they have been very effective in reducing road trauma. In specifically US cultural terms, a 'Crusade for Traffic Safety' was launched in 1954, as Packer relates, 'by media men to help you, a media man, to save lives in traffic' (Packer, 2008: 28). It was led by President Eisenhower and involved a personal pledge to be signed by committed drivers, 'to work through my church, civic, business and labor groups to carry out the official action program for traffic safety' (ibid.: 29). However, Packer makes an important point:

> Reorganizing public opinion is an act of producing a new popular truth and in this case through a form of problematization. By this I mean that a phenomenon becomes recognized as a serious public problem through publicity that identifies it as such and suggests it be dealt with through expert explanation. In essence, it becomes a popular truth insofar as it creates a sense that reality is adequately represented, that something is amiss and must be set right, and, most generally, that there is a form of expertise that can sufficiently come to grips with the situation.
>
> Packer, 2008: 31

In a similar way, a popular truth was created about 'traffic congestion'. Traffic congestion is a sign of the very success of the 'private vehicles on public roads' solution to the desire for mobility. But congestion, a body metaphor, was portrayed as a real cost to the economy of a city, a region or a nation (see Low and Odgers, 2012). The corollary is that the cost of traffic congestion amounts to an economic return on public sector investment in infrastructure (see Box 4.1).

Box 4.1 Calculation of the cost of congestion

Saving travel time is a major part – usually the majority part – of the benefits calculated for infrastructure expenditure designed to reduce traffic congestion. The valuation of the cost to be saved is based on three assumptions: First, time spent in travel is time wasted – time that could otherwise be spent in productive activity. Second, if the 'wasted' travel time could be saved by investment in infrastructure, this economic value is equal to the monetary value of the additional income that the average road user would earn by reallocating travel time saved to income-earning activities. Third, a large part of the cost of congestion in a transport system is derived from the aggregate value of the many small time savings that could be made by all the individuals using the system if the infrastructure allowed something approaching 'free flow' of traffic at optimum speed.

This set of assumptions is problematic for five main reasons.

1. Time spent in travel is not necessarily wasted. Given a seat on a train or tram, many kinds of office work can be done: books, papers or reports can be read, telephone calls made, figures can be calculated, reports written on the laptop (Lyons et al., 2007).
2. The assumption that small portions of time saved by individuals can be aggregated to give a social value is highly questionable. Perhaps an hour extra at home or work, or even half an hour, has some experiential value to the individual, but does ten minutes extra have commensurate, or indeed any value? Might not the individual spend more time in bed or watching TV?
3. In order to calculate the cost of congestion it is necessary to have in mind a transport system that provides free flow against which to compare the existing situation. Congestion is a manifestation of the friction of distance: the obvious fact that it takes time to move from one place to another, but how much time is the minimum one can reasonably expect? How fast should one expect to travel? Very rarely, if ever, is the total cost of providing a transport system that can provide free flow under realistic assumptions of induced traffic entered into benefit-cost calculations involving congestion.
4. The assumption that any new road – i.e. any addition to existing road capacity – will directly result in higher average speeds is based on the proposition that since the new road adds to the road system's engineering capacity it enables an increase in average travel speed, and this increase in road capacity will not have a negative impact on vehicle flow, i.e. the number of vehicles that traverse a specific section of road per hour. Button (1993), however, notes that the speed–flow relationship is an unstable one and that, after the point at which a road's engineering capacity is reached, motorists 'often try and enter the road beyond this volume causing further drops in speed and resulting in the speed–flow relationship turning back on itself' (p. 111). Downs (2004: 102) concludes that, 'Once commuters realize the capacity of specific roads has been increased, they will quickly shift their routes, timing, and

modes of travel by moving to those roads during peak periods, thereby filling up the expanded capacity'. So, expanding the road network capacity, 'does not reduce the intensity of such congestion much in the long run' (ibid.).

5 There is now substantial evidence for the proposition advanced by Zahavi and Talvitie (1980) and Marchetti (1994) that people on average spend a more or less fixed – or constant – amount of time in daily travel – thought to be about an hour. If the speed of possible travel is increased, people travel further rather than reduce the time they spend travelling. Metz (2008: 6) cites data from the UK National Travel Survey to show that average per capita travel time in Britain remained more or less constant between 1970 and 2005, a period which saw enormous investment in new and improved roads. Over the same period the distance travelled increased.

Behind this formulation lies what Sachs (1992: 27) called an 'executive syllogism' of progress:

> What the critics of the automobile saw themselves confronted with in the debates of the time could be called the executive syllogism of competition-driven progress: a) technological development cannot be stopped; (b) escape is not an option, so Germany must take the lead; (c) therefore, we are called upon to support the automobile and its industry with all the means at the state's disposal. ... In the face of so strong a consciousness of national responsibility, the critics became subdued and saw their questions – whether the automobile was necessary at all, and whether its advantages outweighed its disadvantages – grow strangely insignificant.

Sachs is writing about Germany at the beginning of the twentieth century, but the story of economic progress and, as traffic increased, that of the cost of congestion, together with an army of experts calculating that cost, lies at the heart of automobility everywhere. The way the problem was conceived was itself shaped by the technological solution: more and better roads for more and better vehicles. Storylines are tools for legitimating policy and coordinating action through coalitions comprising: 'the ensemble of a set of story-lines, the actors who utter those story-lines and the practices in which this discursive activity is based' (Hajer, 1995: 65). The history of the development of supportive transport 'storylines' has not been much recorded, but an Australian study, through documentary analysis and interviews, has identified the web of rationalization for road building in three Australian State capitals: Sydney, Melbourne and Perth over about 50 years (Curtis and Low, 2012).

To reiterate then, a vision of transport doesn't just appear unheralded on the political scene and immediately capture the public imagination and shape public policy. Nor is it a simple response to a technological opportunity. Rather, there is a process of generation in which a possibility triggers aspirations, tapping values, creating stories of explanation and problematization, shaping interests and epistemic communities, and then adding to technological opportunities – all of which support a system that grows and becomes more and more real so that it may sometimes appear to people in power, who are faced with the need to make decisions, that there is really no alternative (see Figure 4.1).

But of course there are alternatives. A cyclone can only exist while it draws on an environment that supplies it with energy from the heat and moisture of a sea. Once over land it loses its energy source. It continues under its own momentum for a time but it must gradually dissipate. So it is with automobility as a socio-technical structure. The source of energy – cheap fossil fuel – powering automobility, as we pointed out in Chapter 3, is coming to an end. But the dissipation of a cyclone does not mean the end of weather. Many of the same driving forces remain: the desire for mobility, the existence of an economy and society based around the mobility of goods and people, cities transformed by automobility, as we discussed in Chapter 2. If a vision of sustainable transport is to gather momentum and become accepted and influential, a socio-technical process of change will have to occur. That process will have some things in common with the process that generated automobility, but in some ways it will be different. Like automobility there

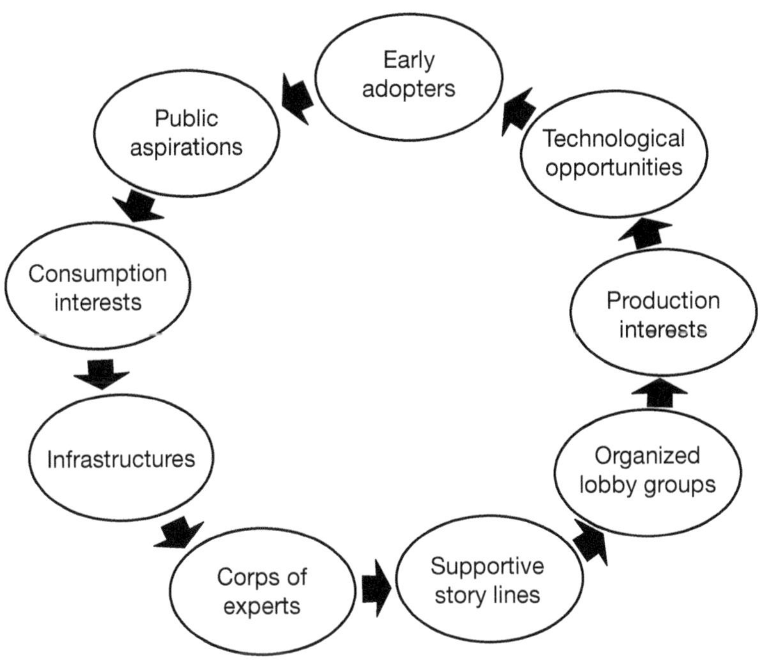

Figure 4.1 The cyclone of automobility (redrawn by Chandra Jayasuriya)

will have to be widespread support of a popular vision. A key difference is that the vision of sustainable transport is not so well aligned with the spirit of market capitalism as was the vision of automobility.

The new vision of sustainable transport

The origins of 'sustainable development'

The idea of sustainable development originated after it began to be understood some 50 years ago that pollution from industrial growth was doing serious damage to the natural environment, and this was not just a matter of local pollution but a global concern. The first images of Earth seen from space powerfully reinforced the idea of a vulnerable planetary environment. But the governments of developing countries meeting at the United Nations General Assembly became concerned that worries in the richer countries about global pollution would harm their own prospects of industrial growth and their chance to generate wealth. So the UN initiated a series of conferences on environment and development, the purpose being to set norms which would steer the world towards forms of industrial growth that would not threaten the quality and integrity of the global natural environment. Hence the Stockholm Conference of 1972 'considered the need for a common outlook and for common principles to inspire and guide the peoples of the world in the preservation and enhancement of the human environment' (UNEP, 1972). The subsequent UN declaration proclaimed that:

> In our time, man's capability to transform his surroundings, if used wisely, can bring to all peoples the benefits of development and the opportunity to enhance the quality of life. Wrongly or heedlessly applied, the same power can do incalculable harm to human beings and the human environment.
>
> Ibid.

The language is anthropocentric. The reports speak of the 'human environment' with 'Man' at the centre. But the underlying idea of 'sustainable development' that came out of these deliberations (including the World Commission on Environment and Development, the Brundtland Commission: WCED, 1987) was that humans can change the nature of industrial development, but they cannot change the limits imposed by the natural world.

However, as consensus was pursued on conflicting aims, the meanings given to 'sustainable development' proliferated. As Sachs (1999: 77) observed, 'Dozens of definitions are being passed around among experts and politicians, because many and diverse interests and visions hide behind the common key idea'. The price for consensus, Sachs observes, was therefore lack of clarity. This lack of clarity allowed 'sustainability' in the language of government to become a green cloak for business as usual. Sustainability, or 'sustainable development' has come to be conventionally defined in terms of a tripartite idea involving the economy, society and environment. While there are many different theoretical formulations of the tripartite idea, the dominant formulation, at least as it appears in public policy, encourages the belief that all three dimensions, economy, society and environment, *must all* be sustained. In lay terms, sustainability therefore now tends to mean that we have to keep the economy thriving, we must hold society together, and we must not deplete the environment too much. In such a formulation sustainable development becomes a political compromise between essentially conflicting ecological, social and economic goals (Low, 2003: 2–4).

As the current global financial crisis unfolds, however, the conflict between business growth, social integration – in Europe – and environmental conservation becomes patently obvious. The escalating price of oil is stimulating a worldwide search for alternative sources of fossil fuel. As Krauss (2011) puts it:

> From the high Arctic waters north of Norway to a shale field in Argentine Patagonia, from the oil sands of Western Canada to deep-water oil prospects off the shores of Angola, giant new oil and natural gas fields are being mined, steamed and drilled with new technologies.

No word about environmental limits there!

Over the 25 years since the report of the World Commission was published it has become much clearer that economic growth – that is growth of business activity – is negatively affecting the global environment. The emission of carbon gases to the atmosphere triggering global warming is only one symptom of the disturbance to natural ecological cycles caused by economic growth (Steffen et al., 2005). As Rifkin (2009: 478) starkly warns, 'Global warming is the entropy bill for the Industrial Revolution'. Despite 'sustainable development' having become a popular mantra in government circles almost everywhere, the world continues on a path of ever greater environmental destruction. We need to start talking about global entropy. Instead of the word 'sustainable' denoting what it was originally intended to mean: economic growth within environmental limits, it has commonly become a way of justifying growth with only minor environmental palliatives. Instead of the environment providing the limiting framework for economic growth, economic growth has become the non-negotiable condition for global environmental conservation.

While the environment could be considered a limitless resource, social and economic development could be thought of as separate matters from environmental conservation. The latter was seen as proceeding from and subject to society and politics, with, for instance, environmentalist groups and Green Parties representing a somewhat marginal social interest in Nature. Now, however, as natural resource limits are approached

– including the capacity to absorb waste – it becomes important to think about the economy as an 'entropic flow': a one way flow through a system of production beginning with resources extracted from the environment and ending in waste put back into the environment. Waste, by definition, is stuff that cannot be reused to create the same level of value as the original resource – CO_2 being a pure example. This is the concept developed by Nicholas Georgescu-Roegen (1971) and further explored by Daly (1996). This concept places the economy firmly within the environment and subject to its limits in a way that, surprisingly perhaps, conventional economics does not entertain.

Daly compares economics with biology. Conventional economic doctrine, he says, 'is exactly as if a biology textbook proposed to study an animal only in terms of its circulatory system, without ever mentioning its digestive tract!' Through the digestive tract, animals 'continuously take in low-entropy matter/energy and give back high entropy matter/energy ... But in economics there is only the circulatory system. This is as true of Marxist economics as it is of neoclassical economics'. He asks, in tutorial mode, 'What concept in economics ties the economy to its environment? Circulation of blood is to circulation of money as the digestive tract is to ... (*what?*)' (Daly, 1996: 193). Climate change resulting from human-generated CO_2 emissions is an irrefutable indicator that an environmental limit has not only been reached, but exceeded. It is not the only such indicator but certainly a very important one for an economy and a transport system heavily dependent on burning fossil fuel. To be meaningful, the term 'sustainability' has to take on board the understanding that the economy, and the society that generates and sustains it – fundamentally, liberal capitalism – is today subordinate to environmental limits.

It should be stated at this point that the above approach to sustainability is compatible with 'multi-criteria' analysis, which assesses policies against environmental, social and economic criteria. Such analysis is legitimate and appropriate just so long as the criteria are kept separated. An example is the work being done in Europe on traction batteries under the SUBAT research programme entitled 'Sustainable Batteries' (Van den Bossche et al, 2005). What is not legitimate, which to its credit the research appears to have avoided despite the misuse of the term 'sustainable', is to bundle all the criteria together and present the bundle as a measure of sustainability. Multi-criteria analysis has the valid aim of finding the most cost-effective, socially fair and technically feasible pathways to an economy that functions within environmental limits. The approach is compatible with Daly's formulation.

Daly's is essentially a prudential approach based on the idea that we have to find practical ways of allowing economic efficiency – so that supply meets demand – and socially fair distribution of wealth to occur within environmental limits. But the idea of sustainability might also be thought of as being imbued with moral purpose: in positive terms conserving the Earth's ecosystems and species evolved over billions of years, or in negative terms avoiding a mass species extinction event (Lynas, 2008: 250). Seen in this light, compromising with sustainability becomes suspect. Do we expect physicians to temper the principle of preserving human health by also supporting the tobacco and alcohol industries? Do we ask lawyers to dilute the principles of justice on the grounds that some interest groups will be hurt by the application of the principles? Do we condone engineers departing from the principle of structural equilibrium when designing a bridge or an engine in order to allow their clients to enhance their profits? The answer of course is 'No'. If unprincipled behaviour sometimes occurs, it is rightly regarded as reprehensible or even corrupt. Seen in this light, those who claim to speak for sustainability are subject to a similar sanction. In view of what we now know about the danger to the world's ecological systems of 'business as usual' economic growth it seems reasonable and natural to think of sustainability as a moral imperative like justice or human health, and for similar reasons. The question of environmental justice in the sense of justice to Nature – ecological justice – is a complex matter which has been explored in depth in the literature of philosophy (see Low and Gleeson, 1998, Chapter 6). Here is not the place to expand on this issue, but the basis of all such moral imperatives is the recognition of the intrinsic value of life: human life, of course, but in the case of environmental justice extending the value of life beyond the human species.

Sustainable transport principles

There are many different values that a transport system should serve other than pure mobility: for instance, public health and safety; economic efficiency; economic development; distributional fairness – or social justice; community integrity; and quality of place – as discussed below. Most are compatible with – and may be supportive of – the value 'sustainability'. But let us call them by their name instead of subsuming all under the one word 'sustainability'. The word 'sustainable' refers to sustaining the global environment.

It is much easier to identify *unsustainable* development than to define sustainability because all human exploitation of the environment necessarily impinges upon it and changes it. It is difficult to specify the limits to economic activity in terms of a broad ideal of sustainability, let alone turn those limits into operational indicators. But it is necessary to do so in order to restore some clarity to the term 'sustainability'. Fortunately both the limits and the indicators with respect to global warming are clear and scientifically established (Stern, 2006; Garnaut, 2008; Pachauri and Reisinger, 2007). So we use those limits as the basis of a simple and unambiguous statement of principle. The first principle of sustainable transport can therefore be stated thus:

- *For an urban transport system to be sustainable it must be one in which the carbon emissions from its operation, and embodied in new infrastructure construction, are reduced to a level compatible with a global temperature rise from pre-industrial levels of no more than 2 degrees Celsius.*

But this principle necessarily has global reach, if only because the atmosphere is a global system, a global commons, and every country's emissions affect the whole system. So we now have to consider a principle of justice as fairness (Rawls, 2001). In such a case it seems reasonable and simple to apply the 'categorical imperative' version of the 'golden rule' enunciated by Immanuel Kant, 'Act only according to that maxim whereby you can at the same time will that it should become a universal law without contradiction' (Kant, 2002 [1798]). The principle finds expression in Article 1 of the Universal Declaration of Human Rights: 'All human beings are born free and equal in dignity and rights' (United Nations, 1948). The Article embodies the concept of equal human rights more generally, including the equal right to property on which the global economy is supposed to be based. Thus the right to pollute the atmosphere in order to generate economic growth is one that applies equally to all people. Starting the application of that principle from the present is not perfectly fair because some nations discovered and followed the path to growth by burning fossil fuel earlier than others, in the process filling the atmosphere with long lived greenhouse gases. Developing nations justifiably grumble that it is the past industrial growth of the developed world that has already loaded the atmosphere with carbon gases. However, retrospective application of a law-like principle of justice comes up against another principle: that people have to know the consequences of their action. Plainly in the case of global warming, the consequences were not known until comparatively recently. These ideas were canvassed in some detail by Baer et al. (2000). Nevertheless, there is an obligation to move towards equality. If then we apply the Kantian imperative, we can state the second principle of sustainable transport thus:

- *For a sustainable urban transport system also to be fair it must be one whose per capita carbon emissions are approximately the same in all cities worldwide.*

These are basic minimum principles of sustainable transport. They are rigorous and will be difficult for rich nations, or rapidly developing nations, to attain. It is important to stress that actions to transform transport almost certainly cannot result rapidly in 'sustainability'. Sustainability is more like a compass point that gives a direction to policy rather than an absolute criterion. But it is more useful for policy to have a meaningful direction than a meaningless criterion. Eventually the direction must become a criterion. The question, of course, is how quickly must this convergence occur? Based on well-established climate science, all transport systems except those in the world's poorest cities are currently unsustainable. The aim for the rich cities of the world, and for the rich socio-economic sectors within them (as we discussed in Chapter 3), must be to move towards zero carbon transport systems within a period of time which would protect the world from dangerous climate change.

Sustainable transport practices

Much more has been written on the practices of sustainable transport – or sustainable mobility – than on the principles these practices are meant to serve. Two recent books on sustainable transport (Schiller et al., 2010; and Black, 2010) include as a condition for sustainability that emissions and waste must be limited to the Earth's capacity to absorb them. Schiller et al. (2010: 2) cite the University of Winnipeg's definition in which the above condition appears to rank equally with such desiderata as the human need for access – presumably to the places people need to go to from home in order to get what they want; human and ecosystem health; equity within and between generations; operational efficiency; and a 'vibrant' economy – vibrating not too much we hope. Black (2010: 4) turns, as we do, to Daly, again noting the sustainability condition: 'the rate of pollution emission does not exceed the assimilative capacity of the environment'. Black says that Daly (1992) does not define what he means by sustainability – though the conditions Black cites surely come close – but in 1996 he does so, specifying sustainability as an economic concept in which the economy is viewed as an open system (Daly, 1996: Ch 2). Daly's analysis is sound and it is this on which we rely to arrive at the basic principles stated above (see Box 4.2).

Having adopted Daly's sustainability condition, and in awareness of the risks of global warming and the IPCC report (2007), Black remains ambivalent about the merits of trying to reduce carbon emissions from transport. Concluding a section on 'Climate Change', he writes:

> Abatement was briefly mentioned, but dismissed as something that does not now seem like a possible strategy for limiting the impacts of climate change. Nevertheless, it does make some sense to try and limit the emissions of greenhouse gases to prevent the impacts from being worse than currently anticipated.
>
> Black, 2010: 35

Exactly what the difference is between 'abatement' and 'mitigation' is unclear. Later he argues that, 'we do not understand the feedback mechanisms at work in terms of the interaction of atmospheric carbon and oceans', or what the resulting increase in temperatures will be. He discusses the impacts of global warming and the attendant uncertainties and writes, 'Until we have a clearer picture of these impacts, incorporating the costs of transport's share of global warming emissions [in the list of costs of automobility] does not appear to have much merit' (ibid.: 92). The consequence of such a view is likely to be continued inaction until it is too late to act to abate global warming whose upper limit will only be reached once the

Box 4.2 Herman Daly's positioning of sustainability as an economic concept

Daly (1996: 48–52) describes the global economy as an open system with inputs – resources – from and outputs – waste – into the environment. He ranks 'sustainability' as a macroeconomic economic concept comparable with 'allocation' and 'distribution'.

Efficient allocation via markets ensures broadly that what people desire – always under the constraint of what they have available to spend – matches what businesses produce. Fair distribution means that there is justice in the distribution of that very capacity to spend. Of course what is social justice in distribution is a contested notion. But governments in democracies today have accepted the task of ensuring that nobody lives in absolute poverty, and that means accepting a concept of fair distribution – or at least limits to unfairness.

In Daly's view, 'sustainability' as a macroeconomic concept relates to the scale of the economy – production, consumption and distribution – in terms of what it takes from and puts into the planetary environment. The important assumption here is that the scale of what is taken from and put into the environment is limited by the environment's capacity to supply inputs and accept waste. Sustainability has come to be understood as an economic concept as we approach a 'full world' – one in which the economy is pressing upon its environmental limits. Global warming is an indicator that one such limit has been exceeded – the limit of a stable climate. Bluntly, nine billion people probably cannot survive on a substantially hotter planet

Daly (op. cit.: 51–52) further distinguishes between a maximum scale and an optimum scale. The optimum scale can be further divided into the 'anthropocentric optimum' and the 'biocentric' optimum. The former is, 'the point at which the marginal benefit to human beings of additional man-made physical capital is just equal to the marginal cost to human beings of sacrificed natural capital'. Natural capital includes environmental 'sinks' absorbing waste. The latter is where 'other species and their habitats are preserved beyond the point necessary to avoid ecological collapse or cumulative decline'. The biocentric optimum is based on recognition that other species have intrinsic value independent of their value to humans.

entire ecology of the planet is transformed. Beyond warming above about 3°C above pre-industrial times, the 'carbon cycle feedback' may begin (Cox et al., 2000). Lynas's explanation of the findings of Cox and his colleagues at the UK Hadley Centre is clear, and worth citing at length.

> Why should the paper, published in *Nature*, have rung alarm bells? First, it showed that global warming could begin to generate its own momentum if a previously unforeseen positive feedback – a vicious circle by which warming would release more greenhouse gas, causing more warming and thereby more gas to be released in an unstoppable

> spiral – came into effect. This, the 'carbon cycle feedback' referred to in the paper's title, would potentially leave human beings as powerless bystanders in a devastating runaway global-warming scenario.
>
> Lynas, 2008: 138

The feedback potential discussed by Cox et al. is not by any means the only paper on the subject. Of course there are uncertainties. That is why climate scientists talk about 'risk'. But at least some of these uncertainties are greatly magnified in the popular media by scientists and others in the service of vested interests in fossil fuel (Oreskes and Conway, 2010).[2] Enough is understood about climate change and published in the scientific literature to indicate that the risks are incalculably large and abatement is the number one issue.

Both Schiller et al. and Black, as well as a host of earlier texts (for instance Banister, 2005), move on fairly quickly, and indeed very usefully, to a variety of problems of *unsustainable* transport and the practices which should be adopted to deal with them. We can only briefly summarize here the quite complex discussions around the kinds of practices advocated under the rubric of sustainable transport.

Public transport

(see Images 4.1.1 to 4.1.19 in the Companion Website)
Most texts on sustainable transport support increased use of collective or public transport – or 'transit' in American terms – for travel both within and between cities (Whitelegg, 1997: 188; Newman and Kenworthy, 1999; OECD, 1995, 2002; ECMT, 2003; Schiller et al., 2010: 96; Humphreys, 2011: 117; and see Chapter 3 above). Glover (2011) posits that public transport is a common pool resource. All of these texts see a shift to public transport as just one among a range of multifaceted policies (33 in Whitelegg, 1997: 204–05). Most couple the necessary mode shift with increased restrictions on car use. Most also discuss the institutional changes required to make the shift possible. In the US context, Black (2010: 270) observes:

> There are some indications that attractive public transit options are proliferating around the United States. If we are ever going to manage the congestion problem, we must have this alternative mode available for use. Little attention has been given to public transit here [in his volume]. In large part this is due to the fact that a doubling of transit capacity would at best double transit ridership. In the United States this progression from 3% to 6% of all trips would not, in and of itself, solve the sustainability problem – but it would contribute to the solution.

We can agree strongly with the second claim – on transit and sustainability – while still questioning the first, for gaining ridership depends on much more than simply capacity, as Vuchic argues (2005); see also Nielsen et al. (2005).

The OECD report on environmentally sustainable transport (OECD, 2002: 96) advocates investment in rail infrastructure to serve both passenger and freight traffic: 'rail movement is inherently more efficient than movement by road, but its stronger feature could be the opportunity it provides for the use of renewable energy sources'. For a public transport system, sustainability depends on the total energy used to move people, as well, of course, as the energy used in the construction of vehicles, together with the source of that energy. Sustainable public transport systems require reasonably high levels of vehicle occupancy, and there can be a difficult trade-off between providing high levels of service throughout the day which can attract users out of their cars, and high levels of vehicle occupancy. But, as the OECD report points out, this does not matter if all the energy used to run the system comes from renewable sources. This is so much easier to accomplish with trains and even buses than with cars. There are also perfectly viable methods of providing efficient public transport services to dispersed – low density – urban environments as described in Chapter 9 of this volume and by Mees (2000, 2010).

A further important characteristic of public transport is that it is much more compatible with a high quality of urban place. For instance, in many European cities, trams, with their predictable – if speed restricted – movements, can mix with pedestrians in walkable spaces – think of the Bahnhofstrasse in Zürich, Switzerland or the Place de la Comédie in Montpellier, France, shown respectively by images 4.1.7 and 4.2.5 on the Companion Website. Even where, for various reasons, there needs to be some separation between trams or buses and pedestrians, the public transport modes use up much less of the public space per person transported than do cars. Nevertheless enthusiasm for public transport needs to be qualified. Maybe, as Hamilton remarks:

> Public transport is not the answer. It will of course have its part to play, but if we are to move on from the individuated, socially divisive and environmentally damaging way of life that accompanies car dependence, then we are simply going to have to travel less.
>
> Hamilton, 2003: 59

And we had better make sure that good public transport opportunities are provided across whole cities and do not become the preserve of the newly affluent, denser inner cores (Schaeffer and Sclar, 2003).

Walking

(see Images 4.2.1 to 4.2.10 in the Companion Website)

Walking is a mode of transport in its own right, not just an ancillary to motorized transport. Tolley's edited book (Tolley, 2003) gathers the work of many writers on sustainability and active transport. The place of walking in sustainable transport is clearly enunciated for instance by

Rees (2003), Hillman (2003), Monheim (2003), and Gehl and Gemzoe (2003) from the varying perspectives of ecology, public policy, transport planning and urban design. They are joined in the volume by many other insightful writers and practitioners. Walking as a transport mode is particularly important in poorer countries where the majority of trips are made on foot, and car travel is a distant aspiration (Simon, 1996; Imran, 2010). As Simon (1996: 95) remarks, 'The vital importance of walking, especially over comparatively short distances, in cities of all sizes is often overlooked. Indeed many people do not even consider it to be a form of transport'. The result is that walking is often mixed dangerously with motorized transport, and walkers are left with no safe space. The same applies even more to people who get around by self-powered wheelchair. Even in rich cities, walkers often vastly outnumber motorists on busy streets but are provided with a much smaller space in which to move. Mees (2010: 184) points out that the neglect of walking as a transport mode leads to underestimation of its importance. He notes that in the journey to work in Zürich, 'virtually all the public transport users walk to the tram, train or bus stop' (ibid.). So the share of walking for at least some part of the journey is 'more like 76%' rather than the 10% recorded as the "walk to work" mode'. So improving the walking environment can encourage public transport use.

Walking is self-evidently the most sustainable mode of transport. Walking, along with sitting, standing and strolling is a social activity. It is essential to the enjoyment of public space and part of every trip, motorized or otherwise. Sustainable transport means reversing the order of priorities: first private vehicle transport, second public transport, and third active transport – with walking at the bottom of the heap (Low et al. 2005: 137). The first priority in cities is to make sure that walking is a safe and enjoyable experience – and to invest public money in infrastructure (streets and squares) to make it so. Kitahara (2003), writing about the small city of Kyojima in Japan, describes the particular qualities that create a walkable neighbourhood: low rise and openness of houses, moderately narrow streets and alleys, places where people come together and socialize. He says continual community engagement and pressure is necessary to protect such neighbourhoods: 'It is necessary to gradually renew a community organization through educating and involving the younger residents in the project' (ibid.: 509). Fenton (2003) also argues that individual interventions to encourage people to walk are not sufficient to stop people reverting to the habit of car use. Rather, collective activities and government interventions to build infrastructure for safe and attractive space, to increase public knowledge and improve public transport are essential to maintain walking for public health. We argue in Chapter 8 that alliances of communities and governments are necessary to transform transport towards sustainability.

Cycling

(see Images 4.3.1 to 4.3.24 and documents 4.3.1 and 4.3.2 in the Companion Website)

As with increased walking, measures to increase cycling for routine trips are part of most programmes for sustainable transport (McClintock, 2002; Tolley, 2003; Horton et al., 2007). Such measures include providing infrastructure specifically designed for cycling – whenever possible separated from motorized traffic – making sure that the cycle path system is a fully connected network without breaks, reducing the speed of motorized vehicles on roads shared by cyclists (PACTS, 1996), providing storage at destinations for bicycles, and combining cycling infrastructure with the public transport network. Some argue that cycling safety is best achieved by having many more cyclists sharing the road with cars. But that will not happen if people feel that sharing the road is unsafe: Catch 22! A plausible indicator of safe cycling environments is the number of women who use them. In the use of vehicles everybody makes mistakes, so as Welleman (2002: 203) argues, 'A truly safe infrastructure will almost automatically lead to the desired form of behaviour of traffic participants'.

The bicycle is possibly the most revolutionary form of transport ever invented. For short distances – 5–7 kilometres – in cities with even minimum congestion it rivals the average speed of cars. It extends the range possible for an average person within 30 minutes from 2.5 to up to 15 kilometres. It can carry surprising amounts of goods. And it does this without a large capital outlay, with almost zero carbon emissions, and with significant benefits to human health – from regular exercise. The cost of providing safe infrastructure for cycling is a small fraction of the amount most countries spend on roads for motor vehicles. With these advantages and a world on the brink of dangerous climate change it is surely negligent not to provide such infrastructure (see Documents 4.3.1 and 4.3.2 for an Australian analysis).

Mees (2010: 189) cautions that an increase in the proportion of journeys made by bicycle within a city or region does not necessarily indicate a step towards sustainability. The question, he writes, should be: 'was this at the expense of the car?' The question is a good one if the proportional use of different transport modes is used to assess sustainability – as it sometimes is. However, the best indicator of sustainability against the Principles is the overall performance of the transport system in terms of GHG emissions. If people choose to cycle rather than take public transport – for instance as seems to be happening in Paris with the Velib system – this will free up space on buses and trains – in Paris the Metro – adding little or nothing to the system's emission load, even though it may not be reducing it. Where public transport systems are overcrowded, with crammed trains, trams and buses, reduced crowding could encourage more people to use them. Also, as Mees acknowledges, cycle networks linked to public transport nodes can assist public transport use in low-density areas, substituting bicycles for cars. This requires good storage facilities at stations for bikes, normally lockable and under cover. Additionally, making places safer for cycling for both recreational and utilitarian purposes is an important part of the broader vision of quality of place associated with sustainable transport, as we discuss below.

Sustainable automobility

The prospect of automobility by other means is extremely attractive to existing business interests and governments because all that seems to need to change is technology. The vehicle and roads industries can remain exactly as they are. The technology for electric cars has been around for at least 120 years (Flink, 1988: 8). Hybrids, hydrogen fuel cells, and alternative – renewable – fuels have been much discussed, and it seems probable that sustainable automobility will eventually play some part in sustainable transport. The potential of such a position has already been discussed in Chapter 3 and little need be added here.

The jury is still out on whether the electric car can be a substitute for the world's internal combustion engine car fleet. At present it seems unlikely that electric cars can become a *sustainable* mass transport mode. The problem is not the electric drive but the storage on the vehicle: the battery, and the energy source of the electricity. Black (2010: 238) considers the possibility that recharging could be done at low demand hours, but acknowledges the uncertainty about whether this could be done without requiring additional power production – and hence increased emissions. We do not yet know whether batteries for electric cars can be produced sustainably or in sufficient numbers to allow substitution on the vast scale of the world's car fleet. Much depends on what they are made of, the availability of the supply of the necessary minerals batteries need, and the emissions cost of resource extraction, manufacture and disposal. There are also major institutional problems. Although some electric vehicle promoters insist that the ultimate source of electricity must be renewable energy, governments will have to legislate to make that compulsory. If left to the market, the cheapest alternative will prevail, and without regulation or incentives that put a high price on carbon emissions, that will remain fossil fuel for many years.

With alternative fuels, for sustainability what one would have to look for is a biological fuel source that does not compete with food production in the use of land. Such a possibility of mass fuel production from such a source perhaps exists in the future – for example oil from algae. How distant a future remains unknown.

Low carbon logistics

Compared with sustainable passenger transport, sustainable freight and logistics occupies a much smaller place in the sustainable transport literature. Nevertheless, as will have been apparent from the discussion in

Chapters 1 and 2, the problems of the impact of freight transport are global in scope, severe and difficult to solve. Whitelegg (1997: 21) noted that, 'Freight transport issues are central to the achievement of sustainable transport', and he devotes a chapter to them as well as 'solutions' (ibid.: 211–19). The solutions he advocates involve reducing freight travel by the substitution of 'near suppliers' for 'far suppliers' and the elimination of 'useless transport' – 'circuitous journeys dictated by the pattern of distribution centres and national/international supply chains'. Whitelegg cites research by Strutynski (1995), Holzapfel (1995) and Böge (1995) that estimates that a 50 per cent reduction in distance travelled by goods could be achieved just by substituting 'near' for 'far' suppliers. Further savings in emissions – in the UK – can be achieved by a staged modal shift:

i utilizing rail/road intermodalism and transferring 50–90 per cent of motorway freight onto the railways – cities with tram systems might consider the use of trams as freight carriers (Rein and Roggenkamp, 1995, with an example from Kassel, Germany);
ii utilizing port and waterway capacity to achieve higher levels of transfer away from road freight;
iii reconstructing logistics and advancing the idea of regional production systems to minimize distance travelled by goods (Whitelegg, 1997: 212).

Unfortunately these practices come up against the neo-liberal model of 'least cost' global economics in which mobility is used to overcome distance, production is sited wherever cheap and appropriately skilled labour is available, and environmental cost is discounted. Strutynski (1995) concedes that a five-fold increase in the cost of road freight transport would be necessary to motivate the shift.

Schiller et al. (2010: Chapter 5) provide an excellent update on sustainable logistics and freight. Though they don't cite Whitelegg's pathbreaking work, essentially the prescriptions are similar though some new ones are added:

- Increasing the energy efficiency of freight vehicles.
- Differentiation of the fleet – using smaller vehicles with smaller engines (three wheelers) or even foot-powered cargo bicycles for final delivery.
- Improved intermodal services and facilities.
- Shifting the truck–train breakeven point in favour of rail transport.
- Reforming ocean shipping to reduce fuel consumption.
- Shifting some airfreight to high-speed rail.
- Making companies responsible for recycling the packaging of the goods they sell.
- Reducing unnecessary freight travel and shortening supply chains through fiscal incentives. (Schiller et al., 2010: 140)

Schiller et al. (2010: 150) recognize that, 'Globalization, with its heavy dependence upon long-distance supply chains, works against sustainable development'. Global business as usual entrenches what Whitelegg has called a 'distance intensive economy'. No one has really explored the social and economic consequences of a fundamental change in the global business model. Yet, if climate change is to be mitigated at all, that model may have to change.

The above set of practices itself goes some way to creating a 'vision' of sustainable transport. But something is missing. If we return to the 'cyclone' of automobility, what sets it in motion, and what still partly underpins it is human desire: for the freedom, flexibility, excitement and delight of the self-contained vehicle on the open road. That is a vision. At the start of automobility, cars were expensive and exclusive toys to expand the lifestyle of the rich. Now almost everyone has a car and, despite the best endeavours of the advertising industry, the real experience of mass ownership and use of cars is often daily drudgery. Yes, rich people can own more expensive and luxurious cars but these are only marginally different from the average and their drivers get stuck in traffic like everyone else. Gadgets and leather seat covers are no substitute for the novelty of speed on the open road. The key here, though, is 'lifestyle' which generates aspiration. Aspiration, as we saw, led within 50 years to mass consumption entrenching a vast range of industries and types of employment. Does 'sustainable transport' contain such a vision that could create aspiration in such a way as to support the public and private investment necessary to realize it?

The outcome of using cars for daily travel was apparent at least 40 years ago. In the 1970s Bendixson (1974: 12) wrote, 'The problems are everywhere the same: the incompatibility of giant highways with decent living conditions, and the liquidation of public transport by an advancing army of cars'. Opinions on the car have changed, but, he continues, 'Government response to this changing opinion has in most cases been slow. Many city engineers still believe there is no alternative to catering for increased car use, and the same view remains well entrenched in all ministries of transport' (ibid.). What happened to cities is that quality of place was subordinated to automobility, especially in the suburbs where most people live. The vision of sustainable transport cannot offer a complete parallel to the vision of automobility, based on a single artifact: the automobile. But it does offer to create or restore something of great value. That can be expressed as 'quality of place' – less tangible than the quality of a thing – the best that city living has to offer. Of course it is a subjective concept, but then 'visions' have to be subjective, appealing at once to emotions and ideas to inspire action. While automobility offered fun and escape from the city to the imagined delights of the countryside, sustainable transport offers to make city life itself delightful. Quality of place is not reducible to sustainability or vice versa, but it is more than just a co-benefit. It is a key part of what the vision of sustainable transport offers to the public.

Of course, there are many possible variants of 'quality of place' limited only by subjective parameters which can probably only be found by practical trial and testing. They should be ultimately determined by the residents and workers of the place in question, which introduces the importance of participation in knowledge formation (discussed in Chapter 11). But still one can imagine an environment whose buildings are mostly of human scale and where the natural world of trees, vegetation and water is allowed to flourish, where the sort of things that people like to do in public space are catered for: shopping, sitting around having a coffee or a beer, and where places of work are not too far from homes. The transport dimension of this kind of place requires a network of paths so that movement by foot power is a pleasant and safe possibility, connecting lower density areas with activity nodes and terminals of the rapid transport network that connects the place to the rest of the wider city by train, light rail (trams), and buses. This is the sort of place where many people seem to want to live today; a place where public space is valued and just as attractive as the space in private homes and gardens. This is a place where people can move between private and public space and feel a sense of collective ownership and pride.

This vision of place was developed by those seeking to reduce 'automobile dependence'. It became the 'new urbanism' which Newman and Kenworthy – themselves early advocates of quality of place – describe as:

> A movement that incorporates the need to expose car-dependent assumptions in town planning rules and fashions; it orients instead around a transit system and attempts to create walking environments through denser, more mixed land use, houses fronting streets with garages behind and other design qualities.
>
> Newman and Kenworthy, 1999: 143

Unfortunately the immensely important vision of quality of place has become clouded by debates about urban density, with the unfortunate implication that sustainable transport requires the wholesale reshaping of cities. We agree with Mees (2010: 53):

> If densities of 400, 240 or even 100 [persons] per hectare are really required to make transport sustainable, then we had better begin searching for a new planet to live on. Built form changes very slowly in aggregate – most of the houses New Yorkers or Melbournians will inhabit in 2050 already exist – but climate change requires urgent action, and the other problems of automobile dependence cannot be put off for decades either.

There are, in fact, mutually indispensible visions: quality of place and sustainable transport, but they do not reduce to one another. They respond to different criteria. Measures of density do not provide the bridge between them. Quality of place is the positive opportunity that is an essential counterpart to the negative risk of global warming and the rising oil price. The vision must offer not only escape from the negatives but also something positive and desired. But still it has to be connected back to the principles of sustainable transport stated above. To do that requires measurement of the carbon impact of urban mobility and the vehicles and infrastructures to support that mobility (Hickman et al., 2010). Cities, in the developed world at least, already have embedded, path dependent, spatial forms and are already equipped with transport systems and infrastructure. Change will occur only gradually, step by step, and new transport systems aligned more strongly to the principles of sustainability will need to evolve out of the old systems already in place. Every city is different in this respect. Every city population has different ideas about quality of place. Questions of quality of place and sustainable transport need to be opened up to the public, because it is the public and not the academy that will make change occur.

Both Newman and Kenworthy (1999) and Whitelegg (1997) point out that there is nothing inevitable about automobility (auto-dependence) and unsustainable transport. Newman and Kenworthy (1999: 129) describe ten myths of the inevitability of automobile dependence. We completely agree with these authors. We capture some of the more compelling 'myths' as self-perpetuating storylines about automobility (Curtis and Low, 2012). It may seem in the chapters to follow that by giving attention to the barriers to sustainable transport we are merely further indulging the myth of inevitability. This is not the case, as we hope to show. Our intention is rather to elucidate the barriers in order to broaden the range of political strategies needed to overcome them. That said, as was apparent in the first section of this book, the dilemma is a formidable one of global dimensions, and if we fail to pay attention to it, and to the political strategies required to resolve it, we may be left with nothing but sustainability as political compromise – with some excellent but very patchy local results, without the global transformation in time to avoid dangerous climate change. The evidence of 'ecosocialisation' is growing in which the world's economic and social behaviour is gradually being adapted to conservation of the biosphere under pressure from the triple social movements for global markets, social protection and ecological conservation (Low, 2002). But is that process moving fast enough? Those who care about sustainable transport should not feel comfortable with the speed of progress.

Sustainable transport in market economies

While Principle 1 above can perhaps represent 'environmental sustainability' in conventional thinking, and Principle 2 'social sustainability', 'economic sustainability' has not only been demoted but has been deliberately omitted. This does not mean that there should be no engagement with businesses with an interest in developing the 'green economy'. On the contrary such engagement is necessary. But far too often 'economic sustainability' means sustaining 'business as usual', resulting in a weak compromise, rather than transforming the economy. Of course economic growth and prosperity of national economies within a global regulatory framework are important goals. But they will automatically be sought by governments of market economies. Sustaining the economy is structured into liberal or social democracies because governments and businesses are locked together in a common interest in economic growth – governments because they are dependent on votes, as voters are dependent on employment provided by companies. Even where there is no electoral democracy – as in China – governments are still dependent on the consent of the people.

Charles Lindblom (1977), the doyen of American pluralist political theorists and co-author with Robert Dahl of the concept of polyarchy, argued persuasively that business – or capital in Marx's terms – occupies a privileged position with respect to government:

> Any government official who understands the requirements of his position and the responsibilities that market-oriented systems throw on businessmen will therefore grant them a privileged position. He does not have to be bribed, duped, or pressured to do so. Nor does he have to be an uncritical admirer of businessmen to do so. He simply understands, as is plain to see, that public affairs in market-oriented systems are in the hands of two groups of leaders, government and business, who must collaborate, and that to make the system work government leadership must often defer to business leadership. Collaboration and deference between the two are at the heart of politics in such systems.
>
> Lindblom, 1977: 175

The problem, Lindblom argues, is that business can in effect veto not only redistributive measures but also 'proposed solutions to collective problems' (Lindblom, 1977: 347). Importantly, the privileged position of business 'permits it to obstruct policies such as those on environmental pollution and decay, energy shortage, inflation and unemployment, and distribution of wealth' (ibid.).

Since 2008, as governments have hurried to spend billions of taxpayers' money propping up banks and the motor industry, Lindblom's analysis has had particular and renewed resonance. Economic growth and sustainable transport are two essentially different goals, and they should be kept conceptually separate. Inevitably the campaign for sustainable transport will be required to show how its practices may be made compatible with economic growth (see for example Jacobs, 1991 and Eckersley, 2004 for useful theoretical discussions on this topic), but that is quite different from building economic growth into sustainable transport as an a priori principle, let alone as the dominant principle.

Technical efficiency is another matter again. In pursuing the principles and programme of sustainable transport it is obviously sensible to adopt practices that can meet the principles at least cost. But that is an instrumental objective subordinate to the principles. It may not be possible to meet the principles in time to be effective. Nazism would not have been defeated if the production of aircraft, ships and tanks had been left to price incentives set by government fiscal policy. Carbon trading may in theory deliver least cost outcomes, but it is also more complex to administer and more open to corruption than a simple tax. Technical efficiency can all too easily displace the principles along the lines of this false syllogism: 'market solutions are the best way to guarantee that least cost solutions are pursued, therefore market solutions are the ones that should always be adopted'.

Automobility was never simply a market-based solution. In fact, the vision was implemented with a great deal of help from government intervention – or 'Fordism' as we discuss in the next chapter. But the vision of automobility is strongly compatible with the disaggregated ideology of market capitalism: individuals finding their own solutions in pursuit of a goal – mobility – organized only by price, within a broad regulatory framework provided by governments. At a micro level, government interventions – for example, providing roads and safety regulations – could be justified by the need to facilitate individual mobility solutions. Governments could look at trends expressed in actual travel behaviour and simply provide for their continuation. Congestion was treated as evidence of demand.

Sustainable transport at its core is very different from automobility in relation to market capitalist societies – whether democratic or otherwise. Sustainable transport, with placemaking and urbanity as its core vision, requires public planning, a concept that has lost favour with the advance of neo-liberalism. The very terms 'public' or 'collective' transport clash with the individualism at the heart of the neo-liberal project. It is no coincidence that many of the paradigmatic examples of placemaking and sustainable transport are to be found in the social democracies of Europe and South America. So, instituting sustainable transport as the dominant mobility paradigm faces considerable barriers. In Chapter 5, we analyse some of the complexities of global regimes of regulation of market capitalism – Fordism and neo-liberalism. In Chapter 6, we examine at a theoretical level, some other institutional barriers to sustainable transport.

Conclusion

In this chapter we have examined a process, discussed sustainability as a political compromise, a prudential economic principle and a moral imperative. Two clear principles of sustainable transport have been posited. Under the rubric of the latter we have summarized some of the practices necessary to implement the principle, and postulated that their achievement will require concerted struggles to overcome barriers. These barriers are formidable. The array of industrial and labour interests in maintaining the status quo is vast. The task of industrial restructuring for sustainable transport is therefore enormous, but has hardly been touched on in the literature to date (but see Weaver, 1998). The introduction of a carbon price, either through a tax or an emissions trading scheme is unlikely by itself to have much impact on travel behaviour or choice of transport mode. The price of fuel will have a larger impact (see Document 4.1 in the Companion Website for an Australian analysis). In response to price increases attractive alternative transport options must be available in order for people's travel behaviour to change.

The process through which automobility became established as the dominant paradigm for mobility cannot be simply replicated for sustainable transport. But understanding processes – socio-technical processes that are deeply political involving social and economic change – can help us to understand what is involved in transformation from one paradigm of mobility to another. Automobility did not occur without aspiration and popular struggle, and those elements will almost certainly have to be present in the transformation to sustainable transport. A broad coalition of interests will have to be constructed around both the positive goals – achieving a better quality of life in urban space – as well as the negative – avoiding threats such as climate change and reduced mobility resulting from escalating fuel prices.

Unlike the struggle for automobility, the struggle for sustainable transport will not find an easy accommodation with the dominant politico-economic ideology. On the contrary, moving to sustainable transport requires a kind of thoughtful, consultative and intelligent public planning which is the antithesis of the 'leave it to the market' doctrine of neo-liberalism. In some places that kind of planning will have to be revived or reinvented as a precondition for sustainable transport. Nevertheless at the time of writing the world seems to be on the cusp of change. The 'global financial crisis' starting in 2008 is only the latest in a series of escalating economic crises, each more severe than the one before. If history is any guide to the future, the incapacity of the dominant ideology to resolve the fissures in capitalism will lead to a moment favourable to transformation. In the next chapter we develop a framework within which to explore such critical junctures for transformation in an East Asian context, that of Japan.

Notes

1 The term 'epistemic community' refers to 'a network of professionals with recognized expertise and competence in a particular domain and an authoritative claim to policy-relevant knowledge within that domain or issue area (Haas, 1992: 3).

2 We hasten to add that Professor Black is not one of them.

5 Capitalist regulation and the provision of public transportation in Japan

Fujio Mizuoka

Introduction

In the preceding chapter a vision of sustainable transport was outlined. One major obstacle to its adoption worldwide was said to be the global regime of neo-liberal capitalism itself. In this chapter, we explore changes in capitalism in the context of Japan, the barriers this context creates for public transportation, and how these barriers have been overcome at times by concerted public action. Here a form of Marxist analysis is employed which went out of favour for a time, with the apparent success of neo-liberal deregulation in the 1990s and its triumph at the collapse of Soviet communism, but which is appropriate in the context of the ongoing economic crisis in the developed world. It is perhaps worth recalling Lindblom's words:

> Even at this late date in the history of social science, we must still turn back to Marx to understand, for example, the adverse effects on democratic government of property rights and of their grossly unequal distribution. Property is a system of authority established by government, just as what we call government is itself a system of authority. In its preoccupation with problems of authority in government, liberal democratic thought remains insensitive to problems of that authority which is embodied in property rights.
>
> Lindblom, 1977: 8

The built environment plays a unique role in the capitalist economy, capturing investment in the physical infrastructure of cities and regions (Harvey, 1982). As argued in Chapter 4, transportation systems constitute an important and integral part of urban built environments, and the urban built environment works properly only as an integrated system of land use – territories – and transport – routes and nodes. The construction and maintenance of transportation systems can share the same properties for business interests as those of the built environment in general, being the recipient of capital investment. So investment in public infrastructure does not simply emerge from the coordinating force of the invisible hand of the market. In a democratic society, such investment can only be made after deliberation supported by sufficient rational argument to persuade taxpayers that the investment is legitimate. As with the other elements of the built environment, the flow of capital takes place through a real political process in which various contending socio-economic classes and social groups are involved.

The classes and groups constituting the society commonly share the same built environment, thus generating a common interest and physical ground for a class alliance (Harvey, 1985). An efficient, user-friendly and socially accepted system of public transportation is thus in the joint interest of business owners and commuting labour, both of which groups need a trouble-free and efficient physical infrastructure to support labour markets, while maintaining the domination by business owners over labour. To achieve such an outcome requires sophisticated politico-economic arbitration and struggle.

In analysing transportation within this socio-economic perspective, therefore, we introduce a conceptual frame of reference that integrates economic and political processes and resultant social structures. It is the theory of capitalist regulation, modified to accommodate neo-liberalism as the successor regime following '*Fordism*'. (Some key terms in regulation theory are explained in the Appendix to this chapter). In particular, this framework seeks to identify 'critical junctures', or the points where the existing 'path' of capitalist regulation can no longer be depended upon to sustain the existing politico-economic system. In this respect the regulationist framework shares a concept – 'critical juncture' – with path dependence theory, which is discussed in the next chapter, though the former analysis is more global in scope with the political regulation of capitalism as a world system in its sights. Since, however, the significance of national and local political struggles is acknowledged in the regulationist framework, the varying outcomes of such struggles change the way in which the global regime of capital accumulation is manifested in different territories. Critical junctures appear as the outcome of political struggles within liberal democracies where governments are subject to forces of the grassroots, including parliamentary democracy and the people's direct movements, which may take either peaceful or violent forms.

In order to substantiate the author's argument, the empirical focus of this chapter is given to public transportation in Japan – a necessity for urban mobility in that country's densely populated urban world.

Regulation theory revisited

The concept of capitalist regulation was first developed by French political economists such as Aglietta (1979) and Boyer (Boyer and Saillard, 2002). They claim that the invisible hand of the market mechanism alone is insufficient to maintain the capitalist mode of production as the primary economic structure of society – in Lindblom's terms, the way in which business and government share authority over employment and production; rather governments must intervene to stabilize the operation of capitalism so that potential conflicts do not erupt to destroy it.

The main argument of regulation theory involves the proposition that the state must intervene in order to contain two fundamental 'natural enemies' of capitalism: the crisis of over-accumulation and the class conflict that could undo capitalism and create a socialist revolution. In regulation theory, a particular politico-economic structure that keeps these two 'natural enemies' simultaneously at bay is called a 'mode of social regulation'. Regulation theory thus recognizes the potential of moments of crisis at which the capitalist mode of production has the potential to collapse, and those currently holding ruling power to lose their dominant position over society at large. Capitalism needs to be permanently supported by some kind of political structure; and a pure laissez-faire system is not possible in capitalism because of its crisis tendencies – as recently exhibited in the proliferating global financial crisis from 2008, and the class fault lines which are beginning to erupt into conflict in Europe.

The mode of social regulation most clearly described by the regulation theorists is what they term *Fordism* – after the Italian political theorist and activist Antonio Gramsci – and its partial breakdown in the 1970s. Although their subsequent argument was sometimes intricate, especially when it dealt with the so-called 'post-Fordist mode of regulation' (Amin, 1995), taking their broad line of argument, the tenets of regulation theory offer a powerful conceptual tool in analysing contemporary capitalism in general. A mode of regulation should be designed to guarantee the long term and obstacle-free functioning of the two essential elements of capitalism: capital accumulation and social integration. Instead of delving into the rather complicated theoretical arguments over post-Fordism, in this chapter *neo-liberalism* is taken to be Fordism's successor mode of regulation, since neo-liberalism is also a product of a politico-economic process, and incorporates the attempts to promote capital accumulation and social integration in ways quite contrasting with those of Fordism.

Fordism

The term 'Fordism' came from Gramsci's reflections on American life and the particular philosophy of Henry Ford, the founder of the mass production automobile industry in the United States (see Gramsci, 1971: 302). Ford achieved high productivity in car manufacture through the introduction of a Taylorist mass production system. Its efficiency allowed him to pay his employees high wages so that they became able to afford cars. 'Ford paid top wages, and early Ford labor practices in addition included bonuses, as well as educational, medical and recreational programs' (Flink, 1988: 120). This system proliferated across sectors of industry producing goods for mass consumption throughout the industrialized world. Thanks to the high profit margins made possible by efficient automated production technology, industrial capitalists became able to pay high wages as well as to offer better working conditions – shorter working hours per day, for example – to the working class. This favourable treatment of workers in turn generated increased effective demand for the products that business produced, especially such durable consumption goods as cars, home electric appliances and houses. Workers did not achieve this condition without struggle, of course. Powerful trade union movements, often acting in national and global alliances, had power to engage in collective bargaining with the capitalists and government for higher wages, job security, better working conditions and a shorter working day. When their demands were not met, organized labour went on strike.

The effective demand for consumer goods promoted sales of industrial goods and abated the risk of the crisis of over-accumulation. The working class enjoyed the fruits of this system, with their homes full of durable consumer goods, leading to satisfaction with their lives and a conservative mindset. Workers came to view themselves as a property-owning class with much to lose should the capitalist mode of production be overthrown. The government protected the rights of organized labour through various labour laws. These measures generated a collective mentality that abated the risk of more militant class struggles, which might have developed into socialist revolution. In this respect it is perhaps worth noting that in a study of 23 rich nations, Wilkinson and Pickett (2010: 52) found that Japan exhibited the lowest level of income inequality, which was also correlated with a fairly high level of public trust.[1] The Fordist regime thus made capitalism more stable and sustainable and realized a 'golden age' of post-war capitalism that lasted essentially until the first oil crisis in 1973.

The role of government in Fordism was also significant in infrastructure provision. Without plentiful public investment to support consumption and capital accumulation, the invested productive capital, as well as deliveries of purchased durable consumer goods, would not function well. The mass-produced goods on the move, or those driving for pleasure, or to get to work would get stuck in traffic congestion. The government also made direct investments in public housing. The massive public investment projects were financed by increasing tax revenues accrued from the growing macro-economy.

For a country that had adopted a democratic political system, affluence and concomitant conservatism among the labour movement meant that those in power were able to attract votes from the working class at elections. Therefore politicians had every incentive to promote Fordist policies under democracy – as long as there was money in the government coffers and profits in the accounts of capitalists with which to co-opt the support of the mass of workers.

Neo-liberalism as Fordism's successor

After the oil crisis, however, it had become increasingly difficult for business and the governments associated with it to be able to afford the Fordist regime. The collapse of the Fordist mode of regulation occurred for two reasons: the fiscal crisis of the national and local states, and the profit squeeze in the corporate – particularly manufacturing – sector. The collapse came because of high production costs due to the high wages and better working conditions demanded by unions, together with increased energy costs, and excess productive capital accumulating more than necessary to satisfy the demand for goods (O'Connor, 1984).

The solution was to cut costs in both the public and private sectors. This meant, for government, a more austere fiscal policy through cutting expenditure on public investments and the introduction of the market principle to the public sector where a political process of democracy and welfare considerations had dominated. For the private sector, the global system termed by Flöbel et al. (1980) the New International Division of Labour (NIDL) offered a solution. The location of productive investment was shifted to peripheral countries with an ample pool of low-wage labour and little in the way of job protection through trade unionism. This 'global shift' not only saved businesses from paying Fordist high wages abroad, but also gave capital the leverage to undermine the collective power of labour at home. It also provided the opportunity for transfer pricing, where profits actually earned in core capitalist countries were sent away to tax havens and peripheral countries where the corporate tax rate was lower.

Ideologically, this policy transformation was endorsed by market fundamentalist thought, which advocated the free market and a shrunken but strong state (Low, 1991: 175–182; Self, 2000: 43–48). This ideology promised that free market competition on a global scale would cultivate a competitive spirit among the privatized agencies and promote trades based on the comparative advantage principle, which would eventually bring about the convergence of various economic variables and egalitarianism among the agencies and macro-economies. The worker was imagined as an independent agent engaging in market competition rather than a member of the working class in solidarity. The powerful unions were either liquidated or co-opted. The struggle for better wages and working conditions, formerly fought as a class, was shifted to individuals competing in the labour market. Collective bargaining for higher wages, job security and better working conditions was replaced by cut-throat competition for employment against one's peers.

These market fundamentalist ideas have their theoretical foundations in the principles of neo-classical economics. In neo-classical economic theory, a unique set of human behaviour principles – having been originally adopted as humble presuppositions in order to develop the neo-classical economic theories on the basis of which to build mathematical models – transformed themselves into the norms of real human behaviour (Takahashi, 2002). Market fundamentalists such as Hayek and Friedman preached to the masses to behave in the way that neo-classical economics presupposes, claiming that such behaviour would be the only road to the blissful millennium where harmonious equilibrium would be brought about by the invisible hand of the market dominating on Earth.

Politicians, initially of the Anglophone world but later almost everywhere worldwide, including Latin America and Japan, enthusiastically supported the turn to the policies advocated by market fundamentalists. They ardently embraced Margaret Thatcher's dictum: 'There Is No Alternative' (TINA) to giving the market free rein, and set off to develop a whole new set of policies that were collectively termed neo-liberalism. The holy trinity of conservative politicians, market-fundamentalist ideologies and neo-classical economists thus dismantled many elements of Fordism, in order to restructure the whole of society and economy. The new capitalist regime of neo-liberalism thus appeared as the successor to Fordism.

Not all regulationists agreed that neo-liberalism was the successor regime to Fordism. For instance, Tickell and Peck did not see neo-liberalism as a new social regime comparable with Fordism with the capacity to resolve, at least for a time, capitalism's internal conflicts. Rather they viewed neo-liberalism as 'a regulatory hole ... which represents the *absence* of a new institutional fix' (Tickell and Peck, 1995: 369). Be that as it may, neo-liberalism survived for more than 30 years, its principles still infuse much public policy at sub-national level, and its core tenets were supposed by apologists for neo-liberalism to have led to the 'Great Moderation' – the taming of the business cycle and a new 'golden age' of stable growth and prosperity for all (see Quiggin, 2010: 10). But capitalism is inherently unstable, capitalist states always seek to contain crisis and revolutionary tendencies, and the 'Great Moderation' was how they proposed to do it following the collapse of Fordism in the 1970s (see Friedman and Friedman, 1980). Among conventional economists, 'disputes over details concealed broad agreement on fundamentals. In particular, few, if any, writers on the Great Moderation suggested that it was approaching an abrupt end' (Quiggin, 2010: 13). Moreover, neo-liberalism has all the characteristics of a distinctive regime strongly propagated by business interests across the globe as the alternative to the Keynesian welfare state. Finally, most of the properties of neo-liberalism are in clear contraposition to Fordism (Table 5.1). Thus neo-liberalism sought a radical annihilation of the old regulatory regime and construction of a new one, eventually leading to a new economic and social crisis from 2008.

Neo-liberal policies and their consequences

Capital, which had accumulated in excess of effective demand, led to the potential for crises caused by overproduction and falling profits. This excess money capital was then channelled to the speculative finance sector, which became the mainstay of capital accumulation under neo-liberal regimes. Neo-liberal governments privatized many public bodies that had previously offered services to the people without the profit motive, which also opened up new fields of investment. Alienating competition was encouraged instead, claiming that it would bring vigour and efficiency to the economy. The notion of the 'post-industrial society' was promulgated, claiming that

Table 5.1 Two contrasting capitalist modes of regulation

	Fordism	*Neo-liberalism*
Capital accumulation	Mass production (Taylor System) – Mass consumption	Gain through financial speculation
Social integration	Conservatism arising from high wages and generous welfare schemes	Competition to join the 'winners' group. Equal opportunity to compete and non-envy the outcome
Staple economic sector	Manufacturing. Finance assisting industrial investment	Finance and information technology. Financialization of industrial investment
Basic sanction	Expulsion from the community	Miserable life for those in the loser's group (for example, fear of absolute poverty and homelessness)
Capital–Labour relations	Strong trade unionism, leading to wage rises, job security, etc., achieved through collective bargaining	Trade unions co-opted to achieve competitiveness. Wage rises through individual competition
Government functions	Income redistribution, managing public enterprises. Social and economic development policies	Controlling market conditions, ensuring competition. Promoting privatization and deregulation. Policing by surveillance
Social ideals	Mutual symbiosis; stable society through 'cradle to grave' government social protection policies	Cut-throat, alienating competition, self-responsibility and the 'no-envy' principle
Social concepts that endorse the regime	Social justice, fraternity, mutual assistance	Market fundamentalism, individualism, self-responsibility
Spatial reach of industrial production	Mainly within the national economy	Globalization based on the New International Division of Labour
The outcome in reality	Relatively egalitarian	Widening discrepancy among socio-economic levels, supremacy of the rich over the poor

people no longer needed to work hard in the dust and dirt of a Taylorist factory in order to earn money, but could gain it on the whim of smart clicks on Internet trading sites. The core countries encouraged financial capital to flourish, and its laissez-faire speculative activities, seeking capital gain across the globe, became rampant.

As for social integration, neo-liberalism adopted the 'North Wind' approach as in Aesop's fable, forcing people to abandon class struggle under the threat of dismal consequences if they did not abide by the market-fundamentalist propositions, as opposed to 'the Sun' approach in Fordism, rewarding people with consumer durables if they gave up class struggle. In neo-liberalism, every person was intimidated to behave according to the assumptions of neo-classical economics, independently as a 'rational economic person' with full responsibility for his/her own decisions. Any consequences, no matter how miserable they may be, must be accepted without envy of those who succeed.

Neo-liberalism penetrated economic policy in the major core countries as the dominant capitalist regime after the 1980s, and the IMF and World Bank demanded that developing countries also accept this regime in exchange for assistance policy packages of 'structural adjustment'. It is important to note, however, that neo-liberalism never managed to gain a sweeping victory in all countries. The People's Republic of China developed a version of liberalism of its own: allowing markets to develop but instead of winding back the state, retaining strong developmental-state type intervention with manufacturing as the macro-economic staple. South Korea also kept some elements of developmental statism and trade unionism.

A little more than two decades after the introduction of neo-liberalism, intense competition generated increasing income disparity between the winners and losers as well as among different social groups and geographical areas. Those economically disadvantaged have suffered from disinvestment in the public sector and welfare provisions, which has caused deterioration of physical infrastructure and of welfare services. Financial capitalism has revealed its inherently unstable nature manifested in repeated grave crashes in spite of carefully designed financial technologies. They erupted in the autumn of 2008, spring of 2010 and again in the autumn of 2011, spreading rapidly from the collapse of housing bubbles to the sovereign debt crises in Europe.

The neo-liberal path of social integration in which social risks were shifted to the individual nevertheless faced repercussions from the masses. In a country under a democratic political system with one-person-one-vote regardless of wealth, neo-liberalism has encountered an inherent paradox. The numbers of the poor that neo-liberalism has produced outnumber the rich by far. Those dissatisfied poor began to resort to democratic political processes to subvert the governments that had imposed the neo-liberal regime upon them, by voting for a political party that represents the poor. This is exactly what has happened in many Latin American countries, some Asian countries, and even in the USA. Faced with the paradox of neo-liberalism, some governments have thus attempted to return to Fordism, and others have sought a middle path with stronger government involvement while still using the market to stimulate the macro-economy. Whether this 'middle path' is a successful solution or just an unstable eclectic grafting of a form of Fordism on to a neo-liberal base remains to be seen.

At any rate, neo-liberalism as a mode of social regulation is itself now on the verge of collapse. The repeated crises of financial capitalism in recent years have deprived neo-liberalism of credibility. It is too early even to speculate on whether a successor regulatory regime will become dominant worldwide, as happened with Fordism and neo-liberalism, or what form it will take. The hardest part is that the financial affluence that would allow the revival of Fordism is no longer available in either the public or the private sectors, and the deterioration of manufacturing sectors in the core countries has eroded the prospect of stable economic growth, which would have generated a solid tax base.

The role of transportation in the regulation of capitalism

In studying the mode of social regulation, conventional regulation economists have mostly focused their research on the process of direct production – how commodities are produced – wage determination and monetary issues while paying little attention to transportation. For example, in a comprehensive list of research programmes in Boyer and Saillard (2002), transportation receives no mention.

But, as discussed in Chapter 2 in this volume, transportation is indeed a crucial factor in making capital accumulation and social integration more stable and efficient. The most fundamental cause of crisis in industrial capitalism is the disruption of supply from demand. Without transportation, relative space stands in the way of the commodity transaction, a physical barrier that has to be overcome between supply and demand (Mizuoka, 2008). Without rapid and efficient transportation with large capacity, the establishments of the multinationals scattered across the globe cannot be well integrated, and the economic network of the NIDL cannot function. Transportation is therefore essential for capital accumulation.

As for social integration, workers must commute every day from their places of residence to places of work in order to sell their labour power. For the most part, workers cannot earn wages, and business cannot find labour without transportation. The leading exception in recent years, of course, is people who work at home with a digital connection to the 'office'. But this is still a small sector of the labour force, and even then 'face-to-face' communication is often necessary and desirable (Scott, 1988: 50–51). Transportation, then, is a fundamental precondition for capitalist social relations.

The means of transportation have been supplied either through the invisible hand of the market or the visible hand of government. Before the Great Depression, the construction of railways was often subject to financial speculation. The market economy thereby facilitated the construction and operation of many railway lines that currently still serve as important transport arteries today. However, after World War II, the laissez-faire market could no longer raise sufficient funds to maintain a transport network sufficient to fulfil the required level of capital accumulation and social integration. Therefore, governments intervened in railway construction and operation, siphoning funds from the market through the 'secondary circuit of capital' (Harvey, 1982).

Transportation in the Fordist mode of regulation

In order to gain the support of labour for the governance of capitalist production, transportation that meets the bare desire for mobility is insufficient. Comfortable as well as efficient transportation, both for commuting and recreational purposes, is essential to co-opt public support. Narrow, unsealed roads or congested trains make commuting and leisure travel quite uncomfortable, generating labour discontent and eventually the disruption of social integration. Under the Fordist mode of regulation governments thus intervened in providing the masses with transportation of better quality.

The efforts of governments were also directed to creating a more homogeneous transport network because politicians saw it as important to integrate everyone, including those living in the remotest parts of the country, into the national transport network. Systems of public transportation formed from the lines and companies originally created by private enterprise came to be owned and operated directly by government agencies – for instance London Transport, and as we discuss below, Japan National Railways.

An important policy principle adopted to achieve a homogeneous railway network was *cross-subsidy*, in which the revenue generated in sectors with heavy traffic was used to offset the deficit generated from sectors in sparsely populated areas, thus money-losing sectors could also be operated without too much financial difficulty. If the cross-subsidy was insufficient, money from taxes was injected to maintain the network – in a general subsidy or, since densely-populated areas pay more tax, another form of cross-subsidy. As long as the sectors with operational deficits were financially supported by these processes of levelling out revenue, a homogeneous public transportation network could be maintained.

However, Fordism supported an important policy agenda that was contrary to the provision of sustainable transport. In order to facilitate capital accumulation, Fordist governments, through regulation, subsidies and road building, enticed people to own and drive private cars, and in so doing provided support for a range of associated industries, thus stabilizing society through a rich endowment of durable consumer goods. Fordist transportation policy did not therefore necessarily prioritize public transportation. It was during the Fordist era, for example, that tramcar networks in many core cities were scrapped in order to make way for smoother driving of private cars. Many inter-city motorways, urban motorways and scenic roads for tourists were built to encourage car ownership (as was discussed in Chapter 4).

Transportation under neo-liberalism

In neo-liberalism, the government is also involved in transport policy, yet in a very different way. The networks of public transportation – homogeneous under Fordism – that had been owned and operated largely by public agencies were split up and privatized, and the management transferred to separate agencies working under the market principle. Under this principle, independent businesses make decisions based on the resources only under their own control. Cross-subsidy is discouraged and each railway line is expected to finance its operations independently of the rest.

Higher speeds – such as in Japan the shinkansen, or in France the TGV – and larger capacities – such as container shipment – became a common strategy to increase the competitiveness of the macro-economy. With the transfer of responsibility for decisions and corresponding existential risks to the individual – without envy for those who succeed – governments under neo-liberalism freed themselves from the obligation of building a homogeneous public transport network for the benefit of everyone.

The practice of cream-skimming became commonplace, where private operators maintained only the most profitable lines, and unprofitable rail lines running in sparsely-populated areas were shut down. In pursuit of more profits even profitable railway lines were the subject of cost-cutting exercises. Investments that did not lead directly to an increase of revenue, for example investments in safety equipment or signalling systems, were often neglected and the lives of passengers put at increased risk. In place of the safe, homogeneous and integrated rail networks available to everybody wherever they lived, the concept of spatial 'free choice' was emphasized. Under this concept (advanced by Tiebout, 1956) individuals bear all the consequences of their choice of residential location. If there is poor transportation, that is also a consequence of individual residential choice.

Nations and cities under neo-liberalism have been encouraged to engage in fierce competition to provide a more competitive business environment, including that of transport networks and hubs. These new modes of transportation tend to be more capital intensive, and cost more per kilometre to build or operate, and thus the transport networks become more unevenly spaced than those of more conventional services. The latter have therefore sometimes been reduced or scrapped as the new services begin to run in parallel. Heavier investment has been made on lines where profit prospects are high, while those generating losses are scrapped without hesitation. Invisible investment in safety and maintenance of tracks, carriages or signalling systems is often neglected as long as fatal accidents do not occur. As a consequence, the egalitarian right to mobility of every member of society is no longer respected as a policy objective, and spatial homogeneity is no longer guaranteed by the 'visible hand' of the government.

Critical junctures and ruptures in the provision of public transportation

No mode of social regulation is without vulnerable moments with the potential to lead to the collapse of the capitalist regime itself. Such moments are 'critical junctures' at which the dominating political group becomes aware of the possible rupture of the mode of social regulation. The 'rupture' occurs when the latent contradiction between the ossified path-dependent structures of the capitalist regime and deteriorating public services and concomitant public dissatisfaction is aggravated beyond the threshold of public tolerance. The 'rupture' in social integration and the 'critical juncture' when it occurs are manifested in a variety of ways: for instance public petitions, protest rallies, political attacks from opposition parties, internet protests or violence of varying intensity, as happened in protests against motorway building in the 1970s and 1980s in Britain, the United States, and Australia (Dudley and Richardson, 2000: Chapter 6; Davison with Yelland, 2004: Chapter 6).

Being so fundamental to capital accumulation and social reproduction, transportation can work as both the trigger of rupture, and also as a tool to contain and limit it. At such moments, if those dominating capitalist society do not want to lose power, they must quench the manifestation of rupture by regaining the consent of the dissatisfied public. Those in power therefore typically use transportation planning and policies to restore stability to the regulatory regime. The 'critical juncture' can therefore play a significant role in the provision of public transportation. The higher the intensity of 'rupture', the stronger becomes the incentive for the dominant social group to restore the regime by improving public transportation. In larger spatial and longer time scales, government attempts may not bring the regime back to the original structure of regulation. The improvement of public transportation takes a considerable amount of political effort, and this effort may introduce new elements foreign to the current regime of regulation, thus changing it. To demonstrate something of the reality of these theoretical ideas, we now turn to case studies from the history of the Japanese railways.

Outcomes for transport in Japan from Fordism and neo-liberalism

Before privatization, provision of a dense and homogeneous network of railways was a statutory obligation. The railway lines were extended or planned to the remotest parts of Japan. Cross-subsidy from the urban service kept the entire network functioning. Trains were operated to cover the nationwide rail network homogeneously. Express trains offered swift connection to most corners of the network, with skilful through service to minimize the need for users to transfer between modes and lines to reach their final destination.

The JNR recorded its most profitable account in 1962. After that, the deteriorating financial position was made up under the Fiscal Investment and Loan Programme by the Japanese government. In 1965 the government, through debt financing, initiated a massive investment programme called 'The Third Long-term Plan' to improve its physical infrastructure via electrification, double-tracking etc. Organized labour in the railway sector, the

JNR and private railways alike, was militant and powerful. It was common in every spring for these unions to go on strike and suspend the rail service for several days in order to achieve a sufficient wage increase. The JNR labourers joined the strikes even though they were officially banned from doing so.

Towards the end of the 1970s, the JNR was facing two knotty issues: the repayment of the massive long-term debt – eventually 27.8 trillion yen – and strong demand from the unions of the JNR to legalize strike action. Under the influence of neo-liberalism, whose principles were adopted by Mrs Thatcher in the UK from 1979, the government began to plan the privatization of JNR as the final solution to this dilemma.[2] In 1982, the Ad Hoc Commission on Administrative Reform of the government in its third report formally proposed the privatization of the JNR. The JNR was privatized in April 1987. Unlike other countries, all the components necessary for passenger rail business – tracks, carriages, operations, marketing, etc. – remained integrated, and the nationwide JNR passenger network was instead divided geographically into six areas, each of which was assigned to a private passenger rail company – JR companies.

All the shares, owned initially by the government, were gradually sold to three Honshu companies, which have eventually become totally privatized. These companies have increasingly been dominated by profit motives. Non-Japanese shareholders were particularly keen in this respect, seeking to shed money-losing rural rail lines and to put more managerial emphasis on the profitable shinkansen and urban networks. Consequently, the investment focus of the privatized JR companies has shifted away from safety equipment that does not directly lead to efficiency, and from infrastructure provision that does not immediately enhance productivity. In some sectors, tight train schedules have been forced onto single tracks or lines with poor safety measures.

The privatized JR companies were freed from the legal responsibility to build a homogeneous nationwide rail network. Prospective rail lines that had been under construction were either abandoned midway or transferred after completion to a public–private partnership (PPP) management, funded mostly by local governments. Many unprofitable branch lines formerly under the JNR ownership were scrapped, while brand new carriages for fast services – 130 km/h maximum on 1.067m gauge track – were provided for the urban network.

Construction of shinkansen that provided higher speed connections only between major cities continued, while express services running direct to rural destinations were abolished. Passengers travelling to rural destinations were canalized to get off at a station of a higher-order regional centre and then drive a car to the destination. The JR promoted this combination of high-speed rail and driving by introducing 'Torenta-Kun' – the train and rent-a-car combo – service at major stations. Rural lines have been intentionally deprived of patronage. Rail lines running in parallel to the shinkansen that were built after privatization have also been transferred to PPPs. After such a transfer, the responsibility for maintaining the rail network has been shifted from the shoulders of government to the people. These lines have been subjected to the campaign for 'my-rail consciousness', or intimidation to 'ride it or submit to scrapping it'. Some lines transferred to public–private management have indeed been scrapped in spite of local protests.

Thus, thinning of the rail network and its uneven development ensued under neo-liberalism. Railways became a means to make money rather than the indispensable infrastructure for the society to ensure mobility.

Critical junctures in Japanese national rail policies in the 1970s

Two case studies are offered below in which at critical junctures there has been a rupture in the mode of social regulation. Since, as geographers have pointed out (for example, Brenner, 2004), the economy and society are organized in multiple layers, it may be that different modes of regulation will become established at different scales – global, national, regional and local. A mode of regulation can be unique to a particular scale. As these case studies show, different interest groups dominate at different spatial scales, seeking different solutions to resolve capital accumulation and social integration.

An urban uprising of dissatisfied railway passengers at the close of Fordism

From the mid-1950s to 1973 Japan enjoyed a 'golden age' of capitalism with a high rate of economic growth. This period essentially corresponds with the Fordist era. There was huge population migration from rural to metropolitan areas where job opportunities abounded. The growth of population in the Tokyo Metropolitan Area pushed its urban fringe rapidly outward (Figure 5.1).

Because of increasing property prices, many commuters could afford only properties along railway lines where frequent city train services were not yet available. Workers were forced to travel ever further between their homes and workplaces on trains that were designed and scheduled for long-distance passenger travel. This limping development of Fordism made for a daily commuting nightmare and generated general dissatisfaction with the JNR, a corporation owned by the national government dominated by the conservative Liberal Democrats.

To put this right, the citizens in urban areas demanded the injection of more funding for Fordist-type public investment and, to that end, elected candidates for municipal socialist regimes at the local level. In 1967, Ryokichi Minobe, a former professor of Marxist economics, won the Governorship of Tokyo with support from the Socialists and Communists, and in 1971, Ryoichi Kuroda, the candidate supported by the same two parties, won the seat of the Governorship of Osaka.

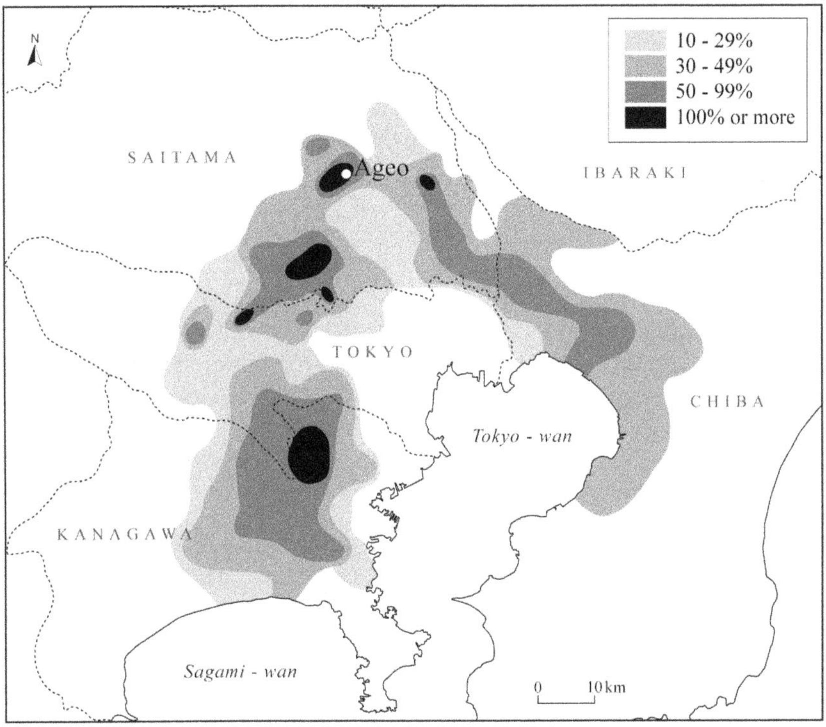

Figure 5.1 Population increase ratio 1965–1970 (redrawn by Chandra Jayasuriya)

Strong trade unionism, especially in the sector of public corporations, was instrumental in this success in 'rescaling' of Fordist politics. The railway sector, both national and private, was the bastion of organized labour in Japan. The nationwide railway system was at that time – late 1960s, early 1970s – still owned and operated by the JNR, a public corporation whose workers were denied the right to strike since they were categorized as quasi-civil servants. Nevertheless, they had several strong unions of varying militancy. The most aggressive was the Rolling Stock Drivers' Union (Doro), the members of which were proud of their role as train operators, maintaining a strong alliance with the union of the British railway workers in the UK (*Asahi Shimbun*, 14 March 1973), from which Doro had learned the tactics of struggle. The largest number of organized JNR workers, mainly the station staffs and train conductors, belonged to the National Railway Workers Union (Kokuro). The most 'tamed' union was the Railwaymen's Union (Tetsuro), which often sided with the railway authority in conflicts with the more militant unions.

Amid this social environment of strong organized labour and radical politics at municipal level in Japan, Fordism quietly approached the moment of its demise. In 1973, the global economy suffered from the first oil shock, pushing up commodity prices for Japanese consumers to the highest level since the hyperinflation immediately after the end of World War II. Due to inflation, real wages had badly deteriorated, which pushed Fordist social integration to the verge of collapse. It was in this context that a violent uprising took place in railway stations in the northern suburbs of metropolitan Tokyo.

Since labour in the nationwide railway system had been deprived of the right to strike, the militant sectors of the national railway trade union movement often resorted to 'work-to-rule' tactics, in which train drivers followed the operation of their rule books to the letter, without regard for keeping trains running on time. Consequently, many trains ran much behind the schedule, which forced commuters to wait at station platforms for long periods until the erratic trains arrived. The events at Ageo station on 13 March 1973 became a critical juncture for the Fordist transport solution.

Events at Ageo Station on 13 March 1973

Ageo, a suburban city in Saitama Prefecture about 40 kilometres from downtown Tokyo, had once been a rural town centre but by the 1960s had been subsumed into the commuting range of Tokyo workers. Its population had doubled in the five-year period between 1968 and 1973 to 135,000 people. Ageo Station was on Takasaki Line, on which the number of train services had increased by a mere 5.3 per cent, from 89 services in 1968 – outbound and inbound combined – to 95 per day in 1973 (*Yomiuri Shimbun*, Saitama edition, 24 December 1999). In 1973, inbound commuter trains ran at roughly seven minute headways during morning peak hours from 6:15 to 8:30am (see Figure 5.2). The number of inbound services had been increased for this peak period by just three services, from 14 in 1964 to 17 in 1973 (*Dai Jikokuhyo*, November 1964). The carriages of some trains had spacious seating catering for long-distance passengers at the expense of capacity. Priority for long-distance express services for distant destinations was also manifested in the sparser half-hour headway during the daytime (*Daiya Esu Jikokuhyo*, 1973).

Shortly after 7.00am on the 13 March, a packed JNR passenger train approached Ageo station. The train originated in the train shed in Kagohara, 71.3 kilometres from Tokyo Central Station and bound for Ueno, a major rail terminus in downtown Tokyo catering for almost all the northbound passenger services. The train arrived in Ageo about half an hour behind schedule, due to the 'work-to-rule' tactics of the militant unions.

The train pulled in to the station in front of around 6,000 passengers who had been standing on the platform and had got tired of waiting (*Asahi Shimbun*, evening edition, 13 March 1973 and morning edition, 14 March 1973). As soon as the train stopped and the doors opened, the passengers, in a hurry to get to their offices on time, rushed into the carriages through the opening doors. However the train was already fully packed, partly because

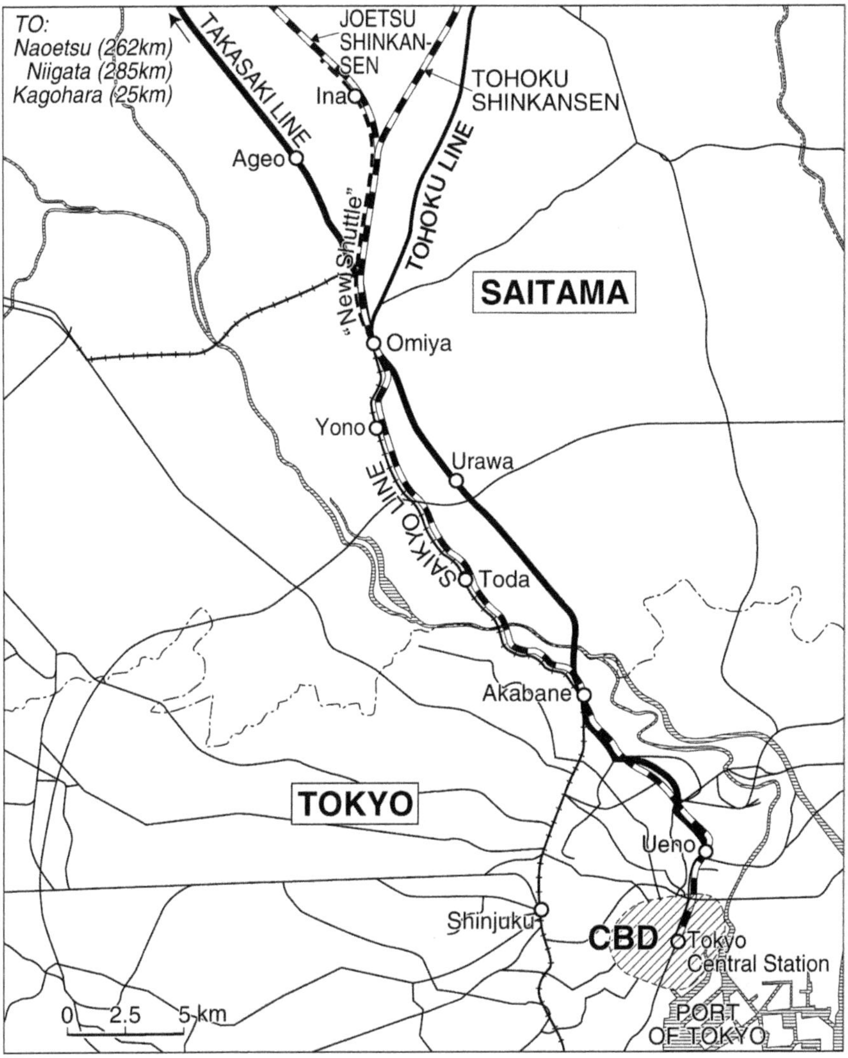

Figure 5.2 The railway route between Ageo and downtown Tokyo

the carriages designed for long-distance travel could not accommodate many people standing. Therefore, not many passengers managed to board the train at Ageo. Soon afterwards another excessively packed train pulled in to Ageo Station, to no avail for most of the passengers waiting to board.

The train was scheduled to turn around at Ueno terminal and become an outbound express for Naoetsu (the 'Myoko No. 2' service). In other words, it was really just forwarding express-train carriages from the suburban train shed to the downtown terminus. The merciful JNR kindly allowed the commuters to board this otherwise unsuitable train. However, having realized that the inbound forwarding train was behind schedule, the Ageo station personnel then suddenly announced that the train would change its destination and terminate at Omiya, an intermediate trunk line station. The intention of the railway management was to minimize the disruption to the schedule of the outbound express by truncating the return trip between Omiya and Ueno.

This decision by the railway management triggered the sudden anger of thousands of passengers already packed onto the train and in the station premises. The passengers set out for violent protest. They broke into the platform operations office, cut off the telephone communication lines, threw stones, destroyed the driving equipment of the trains and attacked the train operators. The protesting passengers also attacked the outbound express train for Niigata approaching Ageo station, as it was the immediate symbol of the oppression of their commuting experiences that had been going on for years. Shortly after 9.00am, the riot police arrived in Ageo station to suppress the uprising. The passengers in revolt then spread out to nearby stations. More long-distance northbound express trains running on the nearby Tohoku Line were also attacked with stones, expressing the passengers' anger at the railway authority giving priority to long-distance train services at the sacrifice of commuters.

Interestingly, there was no evidence of solidarity between the passengers – who were mostly office workers – and the railway station staff and train drivers, even though both groups belonged to the same class. The train drivers and station staff fled as soon as the uprising broke out, turning the deserted station into a kind of 'liberated zone' for the rebellious passengers. The station was simply trashed and robbed of money and unused tickets. The revolt was completely spontaneous, motivated by antagonism towards the railway operations that gave priority to long-distance express trains, and the work-to-rule tactics of the unions. Disorganized, the revolt was crushed after less than ten hours and the train service resumed towards the afternoon and evening of the same day.

However, some Tokyo commuters learned from this Ageo uprising that violence could indeed be a way to have their voices heard by the authorities. When the railway unions resumed their 'work-to-rule' tactics, similar revolts again broke out on the evening of 24 April simultaneously at 26 urban railway commuter stations, mostly along the northbound Tohoku Line. A commuter train was set on fire and burned out, and railway offices were occupied and vandalized by the struggling passengers (*Nihon Keizai Shimbun*, 25 April 1973). Yet again, the unions failed to engage the energy of the rebel passengers in a broadly based movement against the authorities. This might have been a fatal strategic error on the part of the militant railway unions.

The long-term solution in a neo-liberal framework

Faced with the uprising, JNR adopted palliative measures, increasing the frequency of trains a little, and deploying carriages that could accommodate more commuters. The unions took immediate steps to abate the 'work-to-rule' slowdown tactics. Yet, the authorities also started to look for a longer-term solution, beyond the regime of Fordism. Contrary to the will of the unions, this uprising opened up the railway system to the neo-liberal solution. In April 1987, JNR was privatized and disintegrated into seven separate companies.[3] The JNR workers, now employees of a private company, 'achieved' the right to strike. Nevertheless, the privatization of JNR – now Japan Railways, JR – delivered the biggest blow to organized labour, partly because the privatized JR refused to hire 23,660 former JNR employees, including many militant union members of Kokuro and Doro, which had opposed the privatization of the JNR. A group of 1,047 core members of these unions were intimidated and told that if they wanted jobs in JR they should quit their unions. Those who refused were eventually dismissed in 1990 from the JNR Settlement (liquidation) Corporation, which had taken over former JNR employees refused positions by JR (Mataichi, 2010). These core union members kept demanding in vain that JR hire all the former employees of JNR right up until 2011. Consequentially, the union movement in JR was severely undermined, and the unions have lost the real power to go on strike ever since. The 'work-to-rule' slowdown tactic became something of a footnote in railway labour history. Ironically, perhaps, just because of the abandonment of 'work-to-rule' many commuters welcomed this neo-liberal turn in railway administration, at least initially.

In addition, the construction of high-speed railways, another hallmark of neo-liberalism, helped give commuters a better commuter rail service. The Joetsu Shinkansen (bullet train), which had been under construction since 1971 parallel to the Takasaki Line, took over about half of the long-distance passenger services after its completion in 1982. After 1997, the Hokuriku Shinkansen running between Tokyo and Nagano absorbed the rest. After that, almost all the long-distance express services were shifted to the shinkansen tracks; and the conventional Takasaki Line became almost exclusively dedicated to local passenger and freight services. The number of commuter services increased by 144 per cent to 232 between 1987 and 1999, which made commuting more comfortable. (*Yomiuri Shimbun*, 24 Dec 1999). But the costs of heavy investment in bullet trains were yet to be felt as the following case study shows.

Bullet train? No thanks! How the protesting suburbanites of Tokyo secured new commuter rail lines

The means of transportation with greater speed, for example, high-speed railways and motorways, contain an inherent paradox. They can connect distant places in shorter times. But since these means of transportation cost more per kilometre to construct and manage, the network of high-speed transportation becomes inevitably sparser in time and space. Therefore, a more serious problem of spatially uneven development ensues (Mizuoka, 2002: 93–5).

The suburbanites living just to the south of Ageo also engaged in protest against the high-speed rail plan of JNR in the 1970s. The JNR agenda included construction of the bullet train lines southward to downtown Tokyo from Omiya. The commuters demanded that the plan to construct the bullet train be scrapped altogether, because it would not benefit them, and, rather, cost them a worsening living environment caused by noise and vibration of the trains running at high speed.

The plan to build the Tohoku and Joetsu Shinkansen, the bullet train lines radiating northward from Tokyo was approved by the national government in January 1971 (*Asahi Shimbun*, 14 January 1971), and the construction work started in the same year. The residents in the suburbs of metropolitan Tokyo living along the proposed line had not been consulted beforehand by either the national government or JNR. As soon as the plan was announced, they set out for struggle against the project, pointing out the inherent contradiction of the high-speed railway that the bullet train 'forces sacrifice with a hundred harms but does no good [to the residents]' (Mayor of the town of Ina, quoted in Hasegawa et al., 1988: 50). The residents pointed out that the high-speed rail would have no intermediate stations in minor cities or towns to facilitate fast train operations serving those cities, while the citizens' urban space would be split by the elevated shinkansen tracks, and the trains running at high speed would generate noise and vibration. They also feared that some residents, who would be forced to move due to land procurement for the railway right of way, would receive insufficient compensation.

Ina town, a quiet suburb lying approximately 40 kilometres north of metropolitan Tokyo, was to be split physically into three by the 'walls' of the elevated shinkansen tracks. These tracks were planned to fork out in a Y-shape with the north-eastward Tohoku line and the north-westward Joetsu lines right in the middle of the town. The townspeople launched a protest against the project on 11 October 1971, demanding that the route be changed to bypass the territory of their town.

Situated right next to the Tokyo city area, the residents of Toda city, where the shinkansen tracks were planned to run on an elevated causeway for four kilometres before entering a tunnel, were also active for a similar reason: splitting of the territory (Hasegawa et al., 1988: 50). Both Ina and Toda were without easy access to the railways, and the residents either had to use buses or drive to nearby railway stations to catch commuter trains going downtown. The railway that was planned to penetrate their territories would not bring them any benefit by shortening commuting time and distance, but only cause the noise, vibration and separation nuisances they claimed.

A coalition of the local governments was therefore formed on the shinkansen issues in November 1971, and a left-wing candidate critical of the shinkansen project was elected as Governor of Saitama prefecture in July 1972 (Hasegawa et al., 1988: 52–3). The national government and JNR responded to the action on 10 March 1973, offering to investigate the feasibility of constructing commuter rail lines along the shinkansen tracks between Akabane and Omiya, and provision of tram services to Ina town, on the condition that the local communities accept a change in the designs to lay the shinkansen tracks on an elevated causeway through the entire length of Saitama prefecture, instead of placing it underground (Hasegawa et al., 1988: 53). This proposed change of the track structure triggered further struggle against the shinkansen plan by other communities along the route, and the radicalization of the existing anti-shinkansen movement picked up momentum, in the fear of noise and vibration that the trains running on the elevated tracks would generate with concomitant devaluation of residential property (p.55).

In November 1975, Omiya, a city with a major railway node quit the struggle. The reason was obviously that Omiya would have a shinkansen station. Other cities without such favourable geographical positions fought on, continuing to demand that the shinkansen project be scrapped altogether. However, the construction of shinkansen lines continued north of Omiya, and it became clear to everyone that the demand to scrap the high-speed railway project itself was no longer realistic.

In this political environment, the president of JNR announced in August 1977 some tangible conciliatory conditions: first, to respect environmental restrictions – covering noise and vibration – along the shinkansen line; second, to construct a new commuter rail line between Omiya and Akabane along the path of the shinkansen elevated track; and, third, to restrict the speed of the bullet train to 110 kilometres per hour in the section south of Omiya, which would greatly reduce noise and vibration (*Asahi Shimbun*, 10 August 1975). This was in fact the first explicit promise on the part of JNR to construct commuter rail infrastructure between Omiya and Akabane. The governor of Saitama prefecture responded in December 1977, demanding as an additional condition the construction of a tram line running on a track to be built alongside the elevated structure of shinkansen (Hasegawa et al., 1988: 59). JNR accepted this additional condition.

Thanks to these arrangements, the high-speed rail project now offered the benefit of improved commuting conditions to the residents of local communities. This benefit was especially important for Toda, Ina and some other local communities that had previously lacked rapid transit links to the city. In December 1979, the city councils of Toda, Urawa and Yono accepted the plan of the JNR to build the shinkansen on an elevated viaduct (*Asahi Shimbun*, 8 December 1979). The shinkansen now became something that the local residents felt they might endure in exchange for reduced commuting time and a concomitant increase in their property values.

In June 1982, the Tohoku Shinkansen started its service from Omiya to Morioka and in November of the same year Joetsu Shinkansen started operations as well. In March 1985, the shinkansen was extended to Ueno, the northern rail terminus of JNR in Tokyo; and the new commuter rail lines came into service under JNR management in September of the same year in the name of the 'Saikyo (Saitama-Tokyo) line' (see Figure 5.3).

The Saikyo line relieved much of the congestion problem about which the commuters at Ageo station had expressed their rage so violently 13 years before. The tramway, called the 'New Shuttle', began its operation in December 1983 (Funabashi, 1988: 168). Its 12.9 kilometre track northward of Omiya along the Joetsu Shinkansen elevated structure has been operated by a public–private partnership with Saitama prefecture and local community governments as the core investors. In order to promote ridership levels, the prefectural government initiated a plan to build a new town in Ina, with a prefectural social activity centre and a prefectural secondary school that was three times larger in terms of number of students than other prefectural schools of this level (Funabashi, 1988: 175–176).

In the process of coping with this 'critical juncture', the struggles of the citizens worked substantially to change the nature of the high-speed rail investment as a whole: from one that merely connects large cities and nodal stations to one that also caters to the needs of local commuter traffic. The contradiction of spatial integration with high-speed transportation was thus transcended by the power of the residents. Compared with the Ageo uprising, this struggle was more peaceful, deploying more official negotiation channels and lasting much longer. In this case, the struggle generated a critical juncture that might have halted the extension of high-speed rail into Tokyo city centre, which was essential for the increasingly neo-liberal government's plans to integrate the national territory more efficiently. Yet, both cases share the same element of rupture in social regulation

Figure 5.3 JR Saikyo Line (commuter rail) running in parallel to the shinkansen near Musashi Urawa Station (photo by F. Mizuoka, taken June, 2011)

manifested in the strong protest of the residents living along the railway route, a rupture that worked as a catalyst for a better public transportation service and change of regulatory regime.

Conclusion

Transportation is thus integrated into the mode of social regulation in different ways depending on the economic-social scales of space and time that the transport system serves. Each level may have a regulation subsystem that is relatively autonomous to other spatial scales. The national and longer-term scale can be neo-liberal, while local and shorter ones can seek Fordist solutions. Political struggle at a local scale, for example, may induce 'rescaling' of capitalist regulation. The dominating process of regulation may thus shift from one scale to another, at least in some spheres of economy and society. With passenger transportation planning, this 'rescaling' often takes place as the shift between the national and urban scales, as well as long-term and shorter time scales. This is because the range of action space of the masses is normally at an urban scale, which covers the daily commuting range, while the railway is often managed by a national government, seeking longer-term structural adjustment.

A user-friendly, efficient and sustainable public transport system may not emerge through the benevolence of government policy or the market's invisible hand. It sometimes takes a rupture at a 'critical juncture' manifested in citizen protest of various forms and intensities against insufficient or damaging provision of public transportation. The protests must somehow be contained, as they would otherwise escalate and bring about the collapse of the social regulatory regime, especially when the system is more path-dependent and stubbornly unchanging.

As far as transportation is concerned, the containment of rupture can involve heavy capital investment. The dominant political power must be placed at bay as it becomes imperative that it must leave the path that it has hitherto depended upon. We identify here the quantity-quality dialectical nexus between the existing path and the direction of change, with the struggle for transportation functioning as an important element of the nexus. Contradiction becomes rife as quantity augments in the process of development; and at a critical juncture, the existing path transcends itself into a new path of development differently at different scales. In this chapter, this dialectical relationship has been demonstrated, illustrating the dynamism of capitalist regulation in planning and implementing a more sustainable means of public transportation.

Appendix: explanation of technical terms in regulation theory

Some important terms in regulation theory may not be familiar to transport and urban planners. The following provides explanations.

Crisis of over-accumulation

Production in a market economy is based on anarchical decision making by individual capitalists. They produce and supply goods based on the prospect of profit. Since the desire to accumulate money – capital – is without limit, the capitalists tend to accumulate capital without regard for the ceiling of effective demand for what they produce. Full employment empowers workers to institutionalize higher wages limiting wage reduction as a means of protecting profits. Once the market is saturated, a collapse in the market price of products ensues and the rate of profit falls severely. Many capitalist businesses thus fail as a consequence, and the failure spreads through the inputs and outputs of the entire capitalist system. This is the economic crisis of over-accumulation.

Mode of regulation

Economic crisis and socialist revolution are two fundamental 'natural enemies' of capitalism. Economic crisis emerges from the contradiction inherent in a capitalist economy; and the tendency to revolution emerges from the class division and class struggle inherent in capitalist society. To make capitalism sustainable, a capitalist state must produce a social and economic structure to contain these 'natural enemies' by regulating first, the macro-economy through the promotion of capital accumulation, and second, class relations by co-opting the subservient class – labour or sometimes ethnic groups – by paying sufficient wages or providing welfare provisions. To achieve these objectives a unique set of policies is required, which is termed a 'mode of regulation'.

Fiscal crisis of the state

The state takes part in economic activities in a different way from private capital in that decisions on investment and expenditure are regulated by political processes, not by profit motives. Under the Fordist regime, public investment and generous welfare expenditure were supported by excess profits extracted by monopoly capitalists in core capitalist countries. After the 1970s, however, soaring production costs caused by the oil crisis and inelastic wage rates resulted in a profit squeeze. The globalization of capitalism enabled transfer pricing, sending taxable profits away to tax heavens and other peripheral countries. The state failed to collect enough revenue to engage in Fordist economic policies. So the fiscal crisis of the state ensued in which government revenues were insufficient to support Fordist policies, leading to a new neo-liberal mode of regulation.

Taylorism

The term comes from Frederick W. Taylor (1856–1915), an American who devised a theory of 'scientific management'. To make the production process more efficient, the theory promoted an assembly-line production system, where each labourer is asked to perform a single task for the entire length of his/her workday regulated by 'time and motion' studies to speed the process. This approach deprived labour of embodied skill and alienated workers from a sense of participation and pride in manufacturing. However, through the efficiencies that Taylorism achieved, capital could pay higher wages to labour to co-opt workers' consent. Taylorism thus became the fundamental technological base for the Fordist mode of capitalist regulation.

New International Division of Labour

From the late 1970s onwards, the international division of labour shifted from trade in finished products to trade in intermediate products (components of finished products), thanks to three factors:

1 Technological development that disintegrated the manufacturing process vertically and deskilled more matured parts of production technology through Taylorism.
2 Development and deployment of more efficient modes of global spatial integration enabling business establishments of transnational corporations to select optimal locations.
3 Opening up of the massive pool of cheap labour especially in East Asia enabling global capital to cut labour costs.

This is termed the New International Division of Labour (NIDL), which helped capitalists to undermine Fordist labour relations, replacing unionized high-cost labour with docile cheap labour in East Asia.

Secondary circuit of capital

The crisis of over-accumulation could be postponed by siphoning the excessive capital accumulating in the system away from production through the issue of long-term public debt or through taxation. The funds governments thus acquired were then applied to public investment in the built environment, such as motorways, high-speed railways or urban redevelopment, as well as on social infrastructure, such as education or technological research and development, both with long gestation periods. The physical and social infrastructures produced by government in turn served to promote capital accumulation and achieve social integration. The circuit of capital put in motion through government power is termed the 'secondary circuit', as opposed to the 'primary circuit', which passes purely through private capitalist channels. This circuit was commonly employed during the Fordist era, or by countries with a developmental-statist governance model. Under the neo-liberal mode of regulation, excess capital is typically channelled to speculative private investments, and thus the significance of the secondary circuit of capital is weakening.

Notes

1 About 41 per cent of those surveyed agreed with the statement, 'Most people can be trusted'. Data was taken from United Nations Human Development Reports between 2003 and 2006 (Wilkinson and Pickett, 2010: 277).
2 Though in Britain full privatization of British Rail had to wait until Thatcher's successor, John Major, introduced the Railways Bill in 1993.
3 JNR was separated into six geographical regions plus JR Freight.

6 Institutional barriers and opportunities

Sophie Sturup, Nicholas Low, Julie Rudner, Courtney Babb, Crystal Legacy and Carey Curtis

Introduction

If conceptions of how capitalism as a global system is regulated constrain change at a macro level, more local structures, cultures and competing ideas about the ordering of society also influence the development of policy, creating diversity in national and even city regulatory regimes within the broadly neo-liberal framework of ideas. How this diversity in local regimes is generated and maintained in the face of the universalising processes of neo-liberalism can be explained in a number of ways: local path dependencies, professional cultures, definitions of problems, opportunities and risks.

The continuity of public policy from one political administration to the next is something most people take for granted. Root and branch change of policy with each change of government would be chaotic if it were not also impossible. So continuity need not be seen as necessarily a policy failure. Rather this continuity turns into policy failure when the path of policy development becomes inconsistent with the long-term public interest, or when that which is holding matters stable becomes a barrier to publicly desired change. The task of change suggested in Chapter 4 therefore involves a path transition, which not only generates new policies, but challenges those structures which prevent adaptation to new external circumstances – for example, climate change. This task requires an understanding of what holds us to our current path, recognizing opportunities and turning points when they present themselves, and redirecting us to new destinations.

This chapter commences with an analysis of the evidence for path dependence in transport policy using case studies from Australian cities (Low and Astle, 2009; Curtis and Low, 2012). The second section is concerned with forms of action and thought that are constrained by the 'art of government' in play in particular situations. The limitations imposed by specific arts of government are considered through an investigation of mega urban transport projects (Dean, 1999). The third section concerns limitations on the governmental and community capacity to select new pathways imposed by conceptions of risk. Finally we consider governmental capacity to implement policy in terms of negotiations with the public and between agencies and levels.

Working with these concepts means foregrounding the ideas, processes, rationalities, technologies and organizations thar are already present and which act to maintain the current path. Rationalities are generated which constantly reinforce what is already in place. The path to a new sustainable transport system leads not through 'green field' sites of public policy, but rather through policy 'brown fields' which are already densely populated by human mentalities. Change will occur only after sufficient dismantling of what is already present to make room for the new. As was clear from the previous chapter, and as we shall see in Chapter 8, such transformation occurs through the agency of persons and groups.

Path dependence

Research by Curtis and Low (2012) on barriers to sustainable transport uses the 'path dependence' concept as a theoretical framework to explain why transport policy in Australia endures beyond its rational basis and how it could be responsive to change in a geophysical context. Path dependence began in the field of economics to describe aspects of the market which did not to seem to be driven by the rational logic in which supply is driven by demand (Arthur, 1994; Pierson, 2000). It was found rather that, for some products, chance followed by the incentives of real production systems created the momentum to keep producing products even when better ones became available. There are two main claims involved.

1 The presence of a product on the market has a history and can only be understood in a historical context. Circumstances or a confluence of events at the time of initiation of production determine the path taken.
2 Sometimes there is more value to a producer to keep producing essentially the same product – albeit continually improved – rather than change to a different and superior product.

These ideas have been applied to institutional systems that generate public policy by Nobel laureate Douglass North (1990) and political scientists Paul Pierson (2000, 2004) and Jakob Torfing (2001). Consequently in applying the concept of path dependence to transport systems in Australian cities, the research undertook to build an understanding of the historical context in which transport systems arose, and to combine that with an examination of what advantages there may be in maintaining a particular type of system rather than changing it.

Under the theory, path dependence occurs where there are:

- large setup or fixed costs
- learning effects which lead either to better ways to produce the product – reducing costs – and/or consumers learning new and better ways to use the product

- coordination effects so that the benefits to a singular person using the product increase as others adopt it
- adaptive expectations created when it is widely understood that the product will continue to be available and therefore more people take it up.

The research found, through a study of three State capital cities – Sydney, Melbourne and Perth – that these elements are present in Australian transport systems that perpetuate a largely private system of transport based on the motorcar. Clearly there are large setup and fixed costs – notably infrastructure – in the transport system, both for mass transit systems and motor vehicles. There are learning effects: for example the development of private sector engineering firms' expertise in road transport infrastructure design and travel behaviour projection. Learning extends to marketing where the private motorcar has become a symbol of freedom, and to the use of vehicles when transport consumers find new ways to use their vehicles, from travelling further distances through to the development of park and ride schemes. Clearly in the massive public investment involved in the transport system, each individual's capacity to use their car is improved when others take up that mode, because it justifies increasing road building to meet demand evidenced by traffic congestion. So far, governments in Australia have sought to meet that demand rather than allow the experience of the driver to deteriorate as more drivers push road capacities to the point of congestion. Meeting that demand has required massive public expenditure on motorized transport in the form of ever larger and more complex road networks. This, in turn, creates the expectation that such investment will continue and that the motorcar will remain the transport mode of choice.

Thus the current transport system can be characterized as path dependent. This characterization does not, however, reveal much about how this situation came about. Nor does it illuminate what is required to break open the institutional carapace of organizations' policies, beliefs and practices that has been built around the vision of free mobility via the private motorcar. Transport systems and their use result from considerable physical, social, cultural, financial and ideological investments – investments that also serve to protect such systems from external challenges.

Curtis and Low (2012) conclude that the technical dependence of cities on car transport has been shaped by continuous investment in bigger and better roads, which encourages land use patterns serviced by the car. However, the specific circumstances which led to this technical manifestation are complex, and have, among other things, been shaped by a strong, unified, institutional structure for road planning and construction coupled with a weak and fragmented structure for public transport (see also Low and Astle, 2009). This strong and unified organizational structure has been underpinned by a remarkably consistent discourse on the benefits of private motorized transit, a discourse of freedom and modernization. They found that path dependence in the transport system in Australia is thus maintained through three elements:

- the momentum of physical infrastructure
- institutional power determined by well-integrated agencies with the capacity to develop far-reaching policy
- discourse that defines – and thereby limits – the conception of the problem. Policy is shaped by arguments and perspectives, or storylines, supportive of the current direction of policy.

The momentum of physical infrastructure operates in two ways. First, transport choices are conditioned by opportunities, which are embedded in the infrastructure provided. Thus people located in suburbs not serviced or poorly serviced by public transport, but with significant road infrastructure, will travel by car. The assumption of transport modelling is that demand is independent of supply. Yet the now widely accepted postulate of induced traffic implies that supply – of road space – itself generates demand (SACTRA, 1994, 1999; Goodwin, 1996).

Second, the constant production and improvement of the road network perpetuates consumers' belief that the car is here to stay, and thus is the mode of choice for long-term transport needs. Transport and land use become inextricably linked when a high speed road-based network is built, thus enabling urban structuring catering for large car-based catchments with increasing distance between home and work and urban services. This tendency to fractured and dispersed urban form diminishes the possibility of access by other slower modes (see Box 6.1).

Institutions are the framework of rules and norms for policy making. Institutions both limit intentional action and make action possible by providing definitions of problems, solutions to those problems, and the knowledge to implement those solutions. Through organizations they create a corps of personnel bearing that knowledge who can implement solutions. The research has found that in Australian cities, the relative power of transport agencies is unbalanced (Low and Astle, 2009: 50–1). Road agencies have a high degree of control over the full gamut of activities and functions relevant to design, construction and maintenance of the road network. They have direct lines of reporting to both State and federal ministers, and high levels of expenditure and discretional control over that expenditure. In contrast, public transport agencies have traditionally been disenfranchised, with policy, design, construction and operations split between different agencies (some privatized), little direct access to ministers, and a debt burden that cripples their capacity to expand operations without explicit ministerial approval.

Finally, there are significant storylines or supporting discursive propositions that create a shared language for support of road investment in Australia. The research found that there is a high level of agreement about these storylines; even where the conclusions about what should be done

Box 6.1 Technical path dependence

The vast investment in the physical structure of cities itself casts its shadow over the future (Arthur, 1988). Nobel laureate Paul Krugman (1991: 80) observes:

> If there is one area of economics in which path dependence is unmistakable, it is in economic geography – the location of production in space. The long shadow cast by history over location is apparent at all scales, from the smallest to the largest – from the cluster of costume jewellery firms in Providence [USA] to the concentration of 60 million people in the North East Corridor.

The physical form of a city can be shaped by the dependence on a particular form of transport (Schaeffer and Sclar, 1975; Newman and Kenworthy, 1989). In the case of Australian and most North American cities, and the suburban extensions of European cities, the use of the private vehicle for getting around and servicing the city has made possible the choice of low-density residential environments. In turn these low-density environments encourage the building of high-quality roads to accommodate private vehicles, thus reconciling low density with reasonable accessibility to services. (Cervero, 1998: 29) observes for American cities, 'The scattering of activities to all corners of the modern metropolis poses unprecedented challenges to public transportation'. In later work he argues that road investment and urban development are co-dependent (Cervero, 2005: 128–9).

We would argue, however, that such technical path dependence – in the form of urban density – is itself produced by institutional path dependence, and is not a completely determining factor (as also argued by Mees, 2010: 53–5). Once institutional path dependence is overcome, a gradual transformation of transport systems can occur which can be made to serve whatever city form has been laid down over time. We cannot do much to change the shape of cities but much can done quickly to change the institutions governing the creation, management and delivery of transport systems.

in the face of them is challenged. The core road-based transport storyline is one of 'freedom and independent mobility'. The core public transport storyline is one of 'containing the public transport deficit'. (Curtis and Low, 2012: Chapters 8 and 9).

Transport policy in Australia then is a product of the historical growth of the system, and the different capacities of the organizations responsible for developing, building and maintaining it. It is the particular competencies of policy actors that lead to path dependence, not a failure of competence. Accordingly if transport systems in Australian cities are to be taken in a new direction then there is a need to reframe the problem and redirect the competencies. However the research discussed in the next sections of this chapter indicates that this process is far from straightforward.

Art of government

Sturup's investigation into the art of government of mega urban transport projects (MUTPs) was undertaken based on detailed review of three Australian urban mega projects, the Melbourne City Link motorways, the Perth to Mandurah railway and Sydney's Cross City [road] Tunnel (Sturup, 2010 and see Box 6.2).

Mega projects consistently run over time and over budget (Flyvbjerg et al., 2003). This outcome endures despite comprehensive work and

Box 6.2 Case studies of Australian transport mega projects

Melbourne

Melbourne City Link is a 22-kilometre toll road connecting three freeways around the central business district of Melbourne. The road includes upgraded sections of already existing freeway, an elevated section of roadway, a bridge and two tunnels. City Link was built under a Concession Deed granted to a consortium of Transfield/Obayashi to design, build, finance, operate, levy tolls and maintain it for 34 years until 14 June 2034. Ownership will then transfer to the Government of Victoria (Infrastructure Partnerships Australia, 2006). The project cost AUD2.6 billion.

Perth

The Perth to Mandurah railway – also known as the Southern Suburbs Railway or South West Metropolitan Railway – is owned and operated by the State of Western Australia. It stretches 70.1km from central Perth to Mandurah in the south. The line includes two major river crossings, two sets of tunnels, construction of two underground stations as well as numerous above ground stations. Much of the line passes along the central median of the southbound freeway. The operation of the railway is undertaken by Transperth a part of the Western Australian Government's Public Transport Authority. Construction commenced in January 2004 and the line opened on 23 December 2007. The cost of the project is estimated at AUD1.184 billion (Longhurst, 2008).

Sydney

The Sydney Cross City Tunnel project was built under a public–private partnership between the Government of NSW and Cross City Motorway Company. The contract included three stages: design, finance and build twin tunnels – 2.1 kilometres in length following roughly the line of William St; design, finance and build changes to the surface streets – to ensure the decongested streets did not induce further traffic; maintain and operate surface street changes and the tunnel for a concession period of 30 years and 2 months (Catalyst Communications, 2003).

improvement to the key mega project technologies of stakeholder consultation, benefit-cost analysis, engineering, private-public sector contracting, project governance, and project management. Consideration of the art of government of MUTPs was thus an attempt to gain new insight into the underlying conditions that generate and reproduce this outcome.

Following Foucault's theory of governmentality (Foucault, 1991), the art of government of MUTPs can be seen as the amalgam of technological potentials – including the actual products of engineering – the knowledge used in producing them and the rationalities used to justify them, which together provide the ontological reality – or way of being – of the agents within the project, and thus what is possible for them to contemplate doing.[1] Another way of talking about this ontology is that it is the occurrence of the world. Humans act in harmony with the way the world occurs for them. Thus, in MUTPs, actors conduct themselves in harmony with the way the art of government of MUTPs allows the world to occur to them.

This research suggests that the intractability of cost and time overrun in MUTPs is not a product of having the wrong solutions – for example, better costing of externalities in benefit-cost analysis – nor is it a product of the solutions being underdeveloped. Rather, it is suggested that the underlying art of government in which the solutions to these and other problems are developed is different from that in which MUTPs occur. This means that when the solution is used in the MUTP it is adjusted to conform to the MUTP world view. The solution and the MUTP therefore do not 'speak' to each other very well. Within the MUTP, the solutions do not occur as correct, or appropriate, and they are therefore adjusted to support the world view of the MUTP. Sometimes this disjuncture results in solutions seeming oddly incompatible, as though they have been bolted on from the outside as, for example, stakeholder engagement often occurs in mega projects.

Sturup's (2010) analysis found four discernible features of the art of government of MUTPs. The art of government of MUTPs is built from the art of government of 'project' in a more general sense. The art of government of MUTPs is technical. It is pharaonic – or sovereign. And it is organically sticky.

The art of government of project stems from the definition of a project as a unique activity, to be produced within a specific timeframe. In operation it works to ensure that that definition is validated. The technologies of project are designed to maintain the uniqueness of the project and to focus attention on delivery within a timeframe. To achieve this result, the project must be defined within set boundaries of what *is* in the project and what *is not* in the project. Art of government of MUTPs shares these features with art of government of project. MUTPs must therefore be defined as unique activities bounded by time which are also colossal, complex, controversial, captivating, costly, and have control issues (Trapenberg Frick, 2008). An MUTP adds to this definition the construction of some element of the transport system. Like the art of government of project, the art of government of MUTP stems from this expanded definition and similarly operates to ensure that this expanded definition is validated. The control issues referred to by Trapenberg Frick (2008) are a result of needing to place bounds around the activity in order for it to be deliverable within a timeframe. Because of their size and the complexity of their impact, this process of bounding the project is much more complicated for MUTPs than for small projects.

The art of government of MUTPs is in essence technical. Because an MUTP by definition involves the construction of a road, railway, bridge or tunnel, the project is necessarily a physical engineering *solution*. MUTPs are thus technical solutions in search of a problem. If the problem is to be solved by the construction of some physical element of the transport system, essentially by an engineering solution, then the problem needs to be articulated as solvable by that solution. The process of constructing a problem together with its solution is therefore also a process of rendering complex social issues into technical engineering problems that are subject to control. This reification is central to the art of government of MUTPs. Many issues internal to the project are similarly rendered technical issues, such as the allocation of power to act in the project, setting the limits of stakeholder engagement, and ensuring assessment of the project is restricted to whether the project achieved its limited goals such as, for example, delivering the defined object on time.

The art of government of MUTPs always involves sovereign power over decisions. The research showed that all the MUTPs investigated were considered to start at the point at which someone with the capacity to deliver it, promises to do so assuming that the parameters can be satisfactorily worked out. Thus, in projects, power is circumscribed by the technology of project. Power to act in the project becomes a function of the decision to move forward, and is moderated according to the project planning documentation. Individuals in the project are given the power to act by virtue of the decision to go ahead with the project. Thus the allocation of power, and its function in the project, is rendered apolitical and technical. For this process to work, the originating decision must be made by someone who is not only capable of delivering the project, but who is also in a position to decide convincingly that the project should happen. Such an actor also needs to be able to defend this position. The logic of power in the art of government of mega urban transport projects is therefore pharaonic or sovereign – 'I am the king, my will shall prevail'.

Many of the features of MUTPs are designed as a result of, and to guard, this logic. An example is the technology of stakeholder consultation. In MUTPs it is impossible to allow authentic consultation to occur, because such consultation must allow for the possibility that the project should not happen. To question whether the project should occur would unravel both the construction of power within the project and bring into question the power of the decision maker to make sovereign decisions. Since neither outcome is acceptable, the use of authentic stakeholder consultation to alleviate difficulties relating to public protest, which in turn lead to delays and cost overruns, cannot be effectively implemented.

The art of government of MUTPs is organically sticky. In tracking the development of the rationality of the problem and the rationality of the solution of the three projects, it becomes apparent that not all the rationalities deployed in the construction of the problem and its solution are created after the sovereign decision which started the project was made. Indeed the construction of the problem and solution may have started some time before, and makes use of established rationalities in the context in which the project is to be built. For example in Perth, the rationality used to justify the Mandurah project was that which was used to justify the railway extension – the northern suburbs rail line – established after the reopening of the Fremantle line some 20 years earlier. In this sense MUTPs are *sticky* in that rationalities stick to them.

While the rationality of the problem and solution are evolving together, the rationalities used to justify changes to the project, and hence to the understanding of the problem, and why it needs solving, tend to hang around and re-emerge, almost never being entirely abandoned. This finding is not dissimilar to that of Miller and Lessard (2000), who also found that projects evolve. It suggests that like Machiavelli's prince (Machiavelli, 1979 [1532]), the sovereign may not make unilateral decisions simply on a whim. Decisions must be justified in terms of the greater good. In the art of government of MUTPs these sticky rationalities bleed around the decision point, so that absolute demarcation of the project within a particular scope and timeframe is not possible. The result of this stickiness is that even where the decision makes no explicit reference to a particular rationality, that rationality nevertheless sometimes remains a driving factor in the project.

Sturup's work on the art of government of MUTPs provides an alternative mode of analysis to that of path dependence in explaining the stasis of local policy positions. It provides a way of considering the historical development of institutional structure and policy in terms of what world view the rationalities, knowledge and technologies that are used in the project continue to generate in the present. Such world views are the source of being, and define what is logical or 'right' action. Any attempt to adjust the structures that maintain path dependence must be done with consideration for the mode of *being* – ontology – that the new structure will generate. Adjustment will only be successful if the new amalgam of rationality, knowledge and technology is distinguished in such a way that it can be understood as a step change from the world view generated in the past.

Conceptions of risk

Within an overall study about the influence of the conception of risk on children's independent mobility, Rudner (2011) has tapped into another level at which to consider why the transport systems of Australian cities resist change. Together, policy professionals and the lay public create a public 'knowing' of risk that promotes risk aversion rather than exploration and experimentation. Each policy system has a different risk rationality based on the focus of the organization, the problems identified, the methods of data collection employed, and the analyses, interpretations and solutions applied to the issues it faces. Meanwhile there are also differences in risk rationality among experts, and between experts and the lay public.

When we consider transport systems, risk management practices are used to reinforce and maintain current systems, and therefore path dependence. An example is the approach to 'road safety' described in Chapter 4 in which the risks of automobility in the USA were shifted from the transport system to individuals and their behaviour (Packer, 2008: 28–31). Decisions about how to deliver policy outcomes are strongly influenced by technical approaches to risk. Technical approaches are applied from planning and construction to public education on road safety. For example, proven competencies in road building are more likely to pursue low risk projects in road construction than expand into unknown territory such as building new railways. Crash statistics that demonstrate that children are more likely to be involved in road trauma than adults are used to justify control over children rather than control over traffic speeds.

Within many policy and professional domains, risk is considered an externally imposed hazard that impinges on a person or organization; these hazards and their potential consequences are believed to be best evaluated by experts. However, risk assessment experts are trained and socialized within the systems in which they operate, limiting the capacity for organizations to change their approach to risk assessment of projects. For example, Australian road safety auditors consider safety primarily from the 'windscreen' view rather than from the view of the pedestrian or cyclist. To move beyond narrow technical understandings of risk, Rudner suggests an alternative way of conceptualising risk that releases the individual's capacity to look after him or herself. Founded upon an ecological psychology perspective (Gibson, 1986) she provides a pragmatic conception of risk as an interactive relationship between humans and their environment that incorporates notions of agency and competence.

> This re-conceptualization asserts that risk is a relationship between the intent of an agent (individual, group, or organisation), what the environment offers [in the form of opportunities and dangers], and the capabilities of an agent to use the environment in the intended manner. ... Trade-offs between possible costs and benefits in order to potentially achieve desired outcomes are part of ongoing decision-making processes, or stream of action, in the agent–environment relationship.
>
> Rudner, 2011: 20

This new conception of risk is underpinned by the notion of 'affordances'. Affordances comprise the ambient array of objects in the environment that offer action possibilities, and individuals' abilities to perceive and actualize these action possibilities. When individuals actualize perceived affordances, further action possibilities within the environment are revealed, expanding perceptions of affordances and the capabilities to use them (see Box 6.3). However, individuals' use of environmental affordances may be constrained by social and cultural expectations. Through socialization we learn 'what affordances can be utilized, *by whom*, and *when*' (Reed, 1993: 52, emphasis in original). Our actions therefore, are circumscribed by individual 'specializations' such as age, gender, and social status, all of which influence our behaviour, directly and indirectly, although social and cultural expectations may be challenged.

In her study of children's independent mobility, Rudner (2011) found a variety of policies in which assumed hazards and associated risks are assessed as the same for all children. No account is taken of variation among children's different individual capabilities and experiences in relation to the situations they may encounter. Static understandings of risk perpetuate a causal approach – from hazard to victim to damage – that assumes incompetence on the part of children as well as adults and professionals. This associated incompetence arises through a policy focus on precautionary principles, worst possible outcomes, and regulation that defines adult and professional responsibilities and formalizes the education required to comply with policy objectives concerning risk.

Conceptions of risk and how it is managed are a reflection of broader social and cultural norms, expectations and social control, as well as localized differences. Fyhri et al. (2011) found differences in parents' assessments of the dangers children might encounter if they go places by themselves, as

Box 6.3 An example of different affordances arising from different competencies

The 'affordance' includes the particular element of the environment – for example, a staircase – together with the competencies of the person in that environment. For instance, while non-skateboarding, able-bodied individuals may only perceive the affordance of walking up or down stairs, the capabilities and previously actualized affordances of a skilled skateboarder allows them to enlarge the possibilities beyond walking the stairs, to using the skateboard for jumping the stairs, or to slide down the railing.

When affordances are conceived through risk assessment processes, it is possible to say that the trade-off between the possibilities of a successful and an unsuccessful slide down the railing on a skateboard is different for different people. The trade-off is made by the person before doing the sliding, and at that point is only possible or potential. There will be a high risk of a non-skateboarding individual falling off the railing if they try to slide along it. The risk of falling is lower for someone who has rudimentary skills in this trick. For a highly experienced skateboarder, the risk of falling off is much lower than those skateboarders with less experience, or non-skateboarders.

did Rudner and Malone, and Rudner (2011) in their review of research from Australia, South Africa, Japan and Tanzania. For example, more parents from Japan indicated children were less likely to encounter potential hazards if they go places by themselves than in Australia or richer parts of South Africa, and children's independent mobility was more similar to the independent mobility of children in Tanzania (Behrens and Muchaka, 2011; Bwire, 2011; Driandra and Kinoshita, 2011). Essentially conceptions of 'risk' are not universal. Risk to children varies markedly among different countries and at different times. For instance Malone (2012: 13) describes her experience in Tokyo where she encountered groups of kindergarten children walking home from school and crossing busy roads. When she expressed her concern to a Japanese colleague that there was no adult supervision, the colleague responded, 'What do you mean, no adults? There were the car drivers, the shopkeepers, the other pedestrians. The city is full of adults who are taking care of them'. This is a very different attitude to risk and affordance than exists in countries such as the United States and Australia. Where other people are considered to be helpers in Tokyo, elsewhere they may be considered neutral at best or a threat at worst.

In Australia, a single rationality tends to be promoted: risk aversion. Other rationalities that value the costs and benefits of particular activities differently from official policy, or do not view a potential hazard or situation as a risk, are regarded as ignorance that can be rectified with statistical information, rational logic and education. In the case of children's independent mobility, the need for adults to protect children from potential traffic accidents is perceived as appropriate. Whereas children's development, such as the need for exposure to traffic in order to learn self-protective skills, or the need for children to gain a degree of independence, is not considered within risk assessment calculations (see also Chapter 10 of this volume).

While different risk rationalities are viewed as a challenge of behaviour change and management for children's independent mobility, in other areas of transport policy, different rationalities or ways of doing things are viewed as a cost rather than as a behavioural issue. However, both situations are similar with regard to the matter of professional discretion. In matters relating both to children's independent mobility and to transport projects, professionals are viewed as competent insofar as they abide by existing structures and regulations rather than utilize their discretion.

Organizations considering different pathways in transport policy have a tendency to overprice risk in projects that are seen as outside the capacity of the organization concerned, but such projects may nevertheless be within professional capacity. This occurs even though the risks can be greatly reduced by using staff who have appropriate competencies. The result in transport terms is that generally road projects carry a significantly smaller risk premium than rail projects.[2]

Rudner's work points to new ways of understanding the effect of continued investment in road infrastructure, and the degree to which adaptive expectations form our interaction with the world, our conception of risk, and of correct behaviour. Affordances may offer a variety of action possibilities, but how humans relate to action possibilities is enmeshed with social and cultural structures. Both Costall (1995) and Heft (2001) emphasize that knowledge is embedded within the material environment through the discovery and use of tools, the design of artefacts, and the social representations of the material world. The built environment contains the intent of its designers, and the social and cultural intentions of the society. Although individuals may perceive and actualize a number of affordances in the physical environment based on their own objectives and capabilities, they are also guided by the knowledge contained within the material and symbolic landscape. This creates an enlarged public knowing over time and through space between the environment and individuals, a point we explore in the next section. Not only does a continued investment in motorized transport provide increased material opportunities for the use of the car; it also reinforces a cultural commitment to the continued use of the motorcar.

Connecting the public with the planners

One form of resistance to change comes from the broad public. Resistance can result when the policy that the government is trying to implement does not feel 'owned' by those on whom it will impinge – namely the public. While policy ownership is essential to the successful implementation of policy, engaging the community also enables tacit knowledge and people's experience of their city to shape the way we understand the city and the substantive planning challenges it faces.[3] One tool that assists in shaping many cities is the metropolitan strategic plan. Since gaining prominence in Europe – with the 1944 Plan for Greater London among the first – metropolitan plans have become widely regarded for their socio-spatial representation of land use and transport infrastructure, their potential to be either static or dynamic, and their ability to combine social and ecologically based objectives to guide decision making (Albrechts, 2006: 1153; Healey, 1997: 13).

In Australian cities, as is the case in most US cities, such plans guide decision making but often exist without strong legislative and regulatory support. In a policy context where metropolitan plans do not carry significant legislative weight, their influence results from the level of legitimacy achieved through an inclusive and deliberative process of plan making. In seeking to understand how metropolitan strategic plans might achieve greater certainty with respect to implementation, Legacy's research (2010a) drew upon the deliberative democracy literature and on Arendt's (1978) and Benhabib's (1988; 1992) conceptions of enlarged thought to explore the nuanced relationship between process and legitimacy. Enlarged thought means the opening up of different perspectives through a process of open engagement (Benhabib, 1992; see Box 6.4). Inclusiveness and deliberation are highly regarded for their capacity to reduce the incidence of conventionally constrained and predetermined rationalities in determining policy. Drawing on the deliberative democracy literature, an inclusive process promotes ongoing interaction and minimizes power imbalances among stakeholder actors (Habermas, 1990) and can be described as a procedural approach to addressing the range of challenges, including the barriers to implementation. Legacy's study looked at the process of producing enlarged thought in planning the urban environment, but in order to understand the impact that enlarged thought could have on policy decision making, the way in which these processes interface with the more formal institutional planning arrangements was examined. The result was an understanding of the connecting points that bring these two processes together and in particular the unique role that the professional planner plays.

Using metropolitan strategic plan-making processes as the focus of the study, Legacy examined these processes in Greater Perth and Greater Vancouver and paid particular attention to the way the general public, professional planners and politicians were brought together with the aim of negotiating an outcome in the context of the broader planning institutions. The relationship between the planning process and legitimacy was central to this research and was inspired by the need to address the many contemporary challenges confronting cities so that their ecological sustainability is not further compromised.

By engaging with the deliberative democracy literature, the research articulated the importance to the outcome of different ways of knowing. In the process of plan-making, different kinds of knowledge – the community's tacit knowledge and the expert's professional knowledge – are generated

Box 6.4 Enlarged thought

Enlarged thought is a concept arising from the work of Hannah Arendt (1978) and later explored by Seyla Benhabib (1988, 1992). They argue that 'enlarged thought' is the knowledge that emerges from processes that bring people together to reflect on and to gain greater understanding of the many different perspectives that exist on issues. Inclusive dialogue among a range of people, each possessing different sets of knowledge, fundamentally supports the production of enlarged thought: 'the more human perspectives we can bring to bear upon our understanding of a situation, all the more likely are we to recognize its moral relevance or salience' (Benhabib, 1988: 43).

Enlarged thought translates into the opening of different perspectives through some 'dialogic' or 'discursive' ethic (Benhabib, 1992). This 'opening up' is in direct contrast to the kinds of inward reflection and bounded thinking that occurs when the process is restricted to only a few. Enlarged thought requires that the voice of others be expressed (Benhabib, 1992) and is supported through an inclusive process that enables a broader understanding of the challenges, and reduces the impact of instrumental rationality in determining policy.

and exchanged (see Lindblom and Cohen, 1979). By mapping the knowledge exchange over the course of the metropolitan plan-making process it became clear that the knowledge generated is fragmented. Naturally metropolitan plan making is characterized by a lengthy engagement process, and the time needs to be taken to give a range of different actor groups from across the metropolitan region the opportunity to engage with each other at each stage in the process (Legacy, 2010a, b). To retain the deliberative component of the process, each engagement event is usually quite small, and in many cases focuses on the engagement of a single stakeholder group. The research revealed that while the production of knowledge was fragmented, governance arrangements are the ligaments uniting deliberative events with each other, and connecting the broader processes of stakeholder engagement with the formal structures of metropolitan strategic plan making.

Governance is critical to building the legitimacy of the plan as it is the connective tissue between the individual stakeholder engagement events and the broader institution of planning. The critical actor in any governance arrangement supporting a plan-making process is the planner. Not only do planners bring their expert knowledge to the processes, but they too possess tacit knowledge, that is their knowledge of the various barriers to plan implementation that need to be negotiated during the plan-making stage rather than at the implementation stage (Legacy 2010a, b; 2012). A legitimate plan is one that everyone owns (Irazabal, 2005: 52), but it is also a plan that can be implemented.

Legacy's research revealed that substantive impact upon policy, and indeed long-term implementation of policy, is determined not only by how people are engaged, but by the influence that different kinds of knowledge have on policy. The relationship between the knowledge of the public and the knowledge of the professional planner can be reconciled through engagement events that bring these parties together. In the process of developing and implementing a new pathway to sustainable transport, a process will be required to nurture this relationship, and gain the benefits of building legitimacy into the plan.

Government capacity to integrate policy across departments and between levels

Path dependent systems display a large degree of internal coherence. The organizational and discursive institutions that support a particular outcome are highly correlated. This is reflected in the degree of integration of policy – whether explicitly or implied – horizontally across agencies within each level of government, and vertically between levels of government. A constraint on the implementation of a new pathway is, therefore, partly dependent on the ability to create vertical and horizontal integration of new policies sufficient to produce a new transport system. An indication of this capacity can be illustrated by published government policy, both in documents and in the processes through which written policy is enacted.

Policy as expressed in published policy documents, ranging from statutory planning schemes to broad strategic visions, rationalizes planning and decision-making processes within government. Policy in this sense contributes to the framework of norms that guide processes and practices. However, published policy is often a statement of intent or aspiration, and so does not by itself provide an indication of whether governments have the capacity to plan new pathways for transport. The real capacity of policy to address sustainable transport objectives can be restricted by various barriers as published policy is implemented.

In a study of the capacity of government to integrate land use and transport planning in Perth, Western Australia, and Melbourne, Victoria, Curtis, Armstrong and Babb (2010) looked at both these aspects of policy, integration and implementation. An analysis of policy capacity as expressed in published land use and transport documents in Perth and Melbourne was the first stage of the research. In Australia, land use and transport planning is primarily the responsibility of two levels of government – State and local – which have differing powers and geographic scales of interest. Planning undertaken by the State governments at the regional level is an important means to coordinate development, deliver infrastructure and integrate transport networks. State legislation requires local government to create planning schemes that guide the use and development of land within their jurisdiction. Policy capacity in this research was seen as that of integration of policy messages across two dimensions. Vertical integration occurs across differing levels of government while horizontal integration involves policy interactions across sectors within the same level of government. Ideally, land use and transport policy should be well integrated across both the vertical and horizontal dimensions in order to provide the capacity to effect change towards more sustainable modes of transport.

A content analysis of relevant policy documents at the State and local government levels in Perth and Melbourne was undertaken, evaluating policy statements against a comprehensive list of planning principles for land use and transport integration (derived from previous research, Curtis, 2005). Although there was good overall coverage of the principles, integration across State and local government policies was uneven. Vertical integration of policy – between State and local levels – was inconsistent, with gaps in areas such as support for frequent and reliable public transport services and the management of parking. Some criteria that supported effective traffic management were supported in State level documents but had conflicting policy objectives at the local government level. On the other hand there were categories of criteria that were well integrated across the State and local levels, especially in providing for higher density development in close proximity to complementary activities: mixed land use; and the integration of scale and character within precincts. The strength of policy statements at the State level was weaker overall than at the local government level, which, in several cases demonstrated high levels of commitment to the principles. This raises the question of whether the capacity for innovation and responsiveness to local conditions may be compromised when policy messages from the State level, which are prioritized both in their placement in the format of the document and by statute, are weaker than those at the local level.

The second stage of the research sought to investigate policy implementation, and focused on the local government level in Perth, Western Australia. Policy actors within planning, engineering, administrative and political roles at the local government level in Perth were interviewed about various barriers to policy implementation. The research illustrated a range of issues related to the implementation of policy in varying contexts of policy capacity. Local actors pointed to difficulties in how to implement what were considered to be vague State government policy directives. This was especially apparent in urban infill developments where multiple stakeholders were involved. Local government actors considered that State government policy was compromised by a lack of horizontal – or between-department – integration at the State level. Furthermore, the organizational structures of State government, characterized by separate departmental 'silos', increased the complexity and resource intensiveness of consultation and coordination of infrastructure provision.

This research demonstrates that the capacity of government to transition to new models of transport and mobility must be understood in terms of a wider concept of policy that incorporates how it is implemented in diverse contexts. Horizontal and vertical integration of policy across government is a necessary but not sufficient condition for successful uptake of new directions by all actors. It is critical in the process of developing such policy that statements of intent are transformed into implementable action, and that organizational structures are put in place to navigate the complexity and resource intensiveness of consultation and coordination processes.

Conclusion: what then is implied about making change happen?

The work discussed above suggests that there are serious complications in clearing a space for new pathways. There are multiple perspectives that give insight into the resilience of the current policy system. Today's urban transport is the result of interwoven organizations and storylines that constitute the institutions of transport policy. Such interweaving is carried out not only in government policy and structures, but in the stories and technologies used by the community itself. The status quo is not maintained by a simple Machiavellian plot, where those who stand to gain impose their will. The logic of the current system is embedded in society. It is held fast by our understanding of risk, the structures of our organizations, the rationalizations we use to defend decisions and the technologies we privilege. Our conventional understanding of the system, and the technologies and rationalities we use to develop and maintain it presents us with a particular view of what transport is, one which it is difficult to see beyond. Even when we manage to turn our gaze to a new vista and envisage a new system, the myriad of invisible lines pulling us back to the current path reorients us.

Creating a clearing for a new pathway for the transport system takes more than just a clear vision for the future. It will require careful understanding of the complexity of the weave holding the current system in place, and a willingness and ability to reweave the threads to maintain our focus on the new pathway. Perhaps in the final episode we will find ourselves guided mainly by those who can already knit the skeins differently through a different set of affordances. In any case, gaining permission for that future will require a merging of 'expert' and community knowledge through a well-designed consultation process. The shift to a new path is thus a complex task that will take the conscious efforts of many. It will have to be embedded vertically and horizontally not just in government policy and institutions but into the fabric of our cities and the stories that we tell ourselves about transport.

Notes

1 In this sense the Art of Government can be viewed as a particular type of Heideggerian episteme: a way of being which determines what we see and therefore what it is logical to do (Braun and Castree, 1998).

2 Perth's transport planning provides a possible exception where the demonstrated competencies of particular people allowed several rail construction projects to be implemented.

3 Actors bring tacit assumptions, beliefs and habits to the strategic games they engage in. The tacit sphere is the sphere of our thoughts and inter-subjective beliefs (Polanyi, M. 1976).

Part 3
Strategies of transformation

7 Governing dispersed and concentrated cities

Leigh Glover, Carey Curtis, Sun Sheng Han, Haixiao Pan and John Stone

Introduction

The cities of the Asia-Pacific demonstrate extremes of social geography, with the very high density, high rise cities of, for instance, Singapore and Shanghai, and the low density, low rise cities of Australia and New Zealand. Governance systems also vary widely, offering different barriers to and opportunities for sustainable transport. Australian and New Zealand urbanism is most often discussed in relation either to European or North American models. But increasingly delegations from China and other Asian countries are visiting Australia and asking what they can learn from the experience of Australian transport planning. And conversely, high population growth in Australia's major cities has seen its urban planners question the efficacy of car-based suburban growth and look towards Asia for alternative models of development and planning in transport and land use. With an urban population expected to grow rapidly over the next 20 years. Australians are also asking what can be learned from the experience of high-density Asian cities.

In this chapter we compare experiences from Singapore and Shanghai on the one hand, and Melbourne and Perth on the other. In keeping with the integrative philosophy of the book, we find that comparing *unlike* examples leads to new insights – as the Woodcock study has found in comparing London and Delhi (Woodcock et al., 2009). With a common aim – that of reducing carbon emissions from transport – policies have to be designed that are fitted to widely varying local social and governmental circumstances. Global solutions, such as increasing density to reduce fossil fuel consumption – as initially derived from the work of Newman and Kenworthy (1989) – have been seriously challenged (Troy, 1996; Mees, 2010). Yet it is clear that the relationship between land use patterns and transport systems has a significant role to play in moving towards sustainable transport, requiring the integrated management of these two policy domains in 'transit oriented development' (Curtis, Renne and Bertolini, 2009). Here, we turn our attention to a range of sustainable transport problems and issues and examine a variety of public policy responses.

Melbourne: a reckoning of neo-liberal approaches

Australia's second-largest city has achieved wider recognition as an international city by virtue of its regular high ranking in 'world's most liveable city' listings: in 2011 the city was awarded the top ranking in *The Economist* magazine's 'global liveability survey'. Other indices produce less attractive measures, such as the city's very high per capita greenhouse gas (GHG) emissions. This four-million population State capital has a centrally-located CBD surrounded by a ring of inner suburbs dating to the middle and latter nineteenth centuries; beyond this ring are a number of ribbons of suburban development that followed the construction of railways in the late nineteenth century. Between these ribbons lie extensive suburbs largely built in the post-Second World War (WWII) period. While higher density residential development is concentrated in the inner suburbs, the residential density gradient away from the CBD is neither steep nor uniform. Although that part of the city established before WWII is serviced by trains and trams, mobility in the post-war expansion is provided largely by private motor vehicles and sparse bus services. With continual outward growth since WW2, the city now officially covers 7693 sq. kilometres (ABS, 2011).

Melbourne's railway system comprises a network of about 830 kilometres of track on 15 radial lines, with some 212 stations. Its tram system is one of the world's largest, using some 250 kilometres of double track on 29 routes – including a tourist loop – with 1774 stops; most of the tracks are located on public roads. Although it carries fewer passengers than either rail or trams, the bus system services the most extensive area with some 6254 route-kilometres on 337 routes and over 18,000 stops, involving 35 private operators, but with typically much lower service levels than the other modes. Despite these services, Melbourne is a car-dominated city; for example, 78 per cent of the journey-to-work in 2006 was by car, with 14 per cent by public transport, and less than 5 per cent walking or cycling.

There are many transport and land use challenges, but several stand out: a high degree of reliance on private motor vehicles for passenger transport; a high proportion of freight moved by motor vehicles; poor access to transport services by socially and economically disadvantaged groups who are becoming concentrated in locations with poor public transport; population growth leading to urban expansion with few public transport services, leading to high and continually growing environmental and social costs arising from the transport sector. GHG emissions data is unavailable for Melbourne specifically, but based on the most recent Victorian State GHG data – for 2009 – and apportioning this according to Melbourne's population share of the State, we can estimate Melbourne's transport emissions at 14.3 MtCO_2e (ABS, 2011; DCCEE, 2011). On a per capita basis, this amounts to 3.6 tonnes CO_2e. Against the baseline of 1990, the 2009 State emissions represent an increase of 18 per cent (DCCEE, 2011).

Government in Australia is essentially the Westminster model. Under Australia's three-tier system of elected governments – national, State, and local – the governing of the capital cities – such as Melbourne, Sydney and Perth – sits awkwardly. Australia's capital cities in its six States and two territories dominate their territories, indeed, as a group they comprise almost 80 per cent of the national population. All these capital cities – except for Brisbane in the State of Queensland – are without their own governments; instead, they are governed by their respective State and territorial governments. Significantly, federal governments have considerable influence over the capital cities through the exercise of fiscal allocations in tied grants and funding for special projects. A State Department of Transport implements public policy for the road system and for public transport and a new – in 2012 – agency, Public Transport Victoria, oversees all public transport. While the roads system is largely a public asset – although there are some private tolled motorways – the public transport system is operated under franchise agreements with a suburban train operator, tram operator, and several dozen bus companies. Land planning is under the State Department of Planning and Community Development.

After experimentation with corporatization and other reforms, Melbourne's public transport system was privatized on a franchise model in the early 1990s, making it one of the pioneers in this policy realm. Institutionally, the current system is far removed from the colonial bureaucracies that controlled freight, passenger, city and rural public transport, yet vestiges of these older systems remain.

Mixed blessings of public transport privatization

Melbourne's public transport system of trains and trams, and some bus lines, operated under post-colonial State bureaucracies for almost a century – while the majority of its buses have been privately owned, subject to government regulation. Following historical declines in ridership, operational efficiency, and condition of the rolling stock, increases in union protests and disputes, and operating losses since the 1950s, a progressive State Labor government (1982–1992) sought to act. Public transport had proved a difficult portfolio for the government: usage was falling to historic lows and some even considered that its primary function was a welfare service for those without access to motor cars. In response, the Cain/Kirner (Labor) government initiated reform on several fronts: corporatization, greater government control, increased investment, efficiency increases, and workforce reductions. Not all of the intended benefits materialized and there was increased industrial unrest from the strongly unionized workforce, which traditionally supported the Labor Party (Stone, 2009a). Partly as an outcome of these industrial actions – notably a major tram strike – and broader economic problems with the State's role in the economy, the Labor government lost office to a conservative Coalition in 1992.

Following a strongly neo-liberal ideology, the Kennett government (1992–1999) moved significantly to reduce the public sector workforce and began a wide programme of privatization of public agencies and deregulation in what would become the most sweeping changes to the public sector since Federation. Despite an early belief that public transport wasn't an obvious candidate for privatization, by its second term in early 1997, the government began to privatize the State's public transport. Adding to the novelty of these changes was a highly unconventional design for the new system, which divided the tramway system and the rail system into separate operating territories, thereby creating four separate service providers. A franchise system was used in which competitive bidding took place for the 10–15 year contracts; the new system began in mid-1999.

By 2002, all the franchisees were in financial trouble. Several causes were involved, notably the franchisees' expectations of increased ridership – with associated performance bonuses – and of cost reductions that were unachievable. Holders of one tram and one train franchise, the National Express Group, handed back their franchises. All four franchises went into State government receivership for the period 2002–04. During this time, the government revised the franchise model back into one tram franchise to a Transdev/Transfield consortium and one train franchise to Connex/Veolia and let new five-year franchises to Yarra Trams for the trams and Connex Melbourne for the railways, together with large one-off cash payments. Subsequently, a second round of franchises was let in 2009 to new franchisees for eight years to KDR Melbourne (trams) and MTR (trains).

Opinions differ greatly over whether privatization has been a success – and indeed over what constitutes success (Allsop, 2007; Auditor-General, 2005; Mees, 2002, 2005; Stone, 2010). Privatization was predicated on the belief that net public costs would be reduced. Supporters of privatization conceded that the system costs have not been reduced, while critics point to evidence that post-privatization costs have increased, including the State Auditor-General (2005). Expectations that a privatized system would eventually run without public subsidy proved fanciful and misleading; this did not and will not occur. Increases in public transport ridership have been considerable in recent years, but according to the State authorities these are due to factors other than privatization – such as rising costs of motor fuel, increased charges for CBD parking, and job and housing increases in the public-transport serviced areas.

Throughout the era of privatization, both the tram and train systems have been plagued by continual problems, including overcrowding, increasing public costs (in the form of subsidies to the franchisees), rolling stock failures, poor reliability and poor punctuality. Most of the original claims and expectations of privatization have not been met and the portfolio continues to be a major political problem for incumbent governments (Glover, 2010). However, privatization appears entrenched at this time, with neither the Government nor the Opposition entertaining any plans for bringing the sector back into the public realm – noting that the infrastructure and rolling stock have always remained in public hands. Tellingly, no other State or territorial government in Australia has followed the Victorian experiment.

Environmentally unsustainable outward urban growth

Melbourne's rapid growth into a city occurred in the latter part of the nineteenth century and was coincident with the introduction of mass public transport, which then began to play a major role in Melbourne's urban development through to WW2. These fixed rail services are arranged in a radial pattern emanating from the CBD, providing strong corridor access around the city centre and quite limited access between corridors, a task somewhat met through the bus services. In the early 1950s there began a rapid shift to mass mobility based on private car travel, a development that was tied to land use and transport planning. Notably, most of the tram and train lines were in place by the 1930s and there have been very few additions in the following 80 or so years.

Given that the general growth of the city has been outward, although biased to the south and east, the effects of mobility choices are clearly expressed in the city's urban geography. There is an inner city, comprising the oldest settlement, and middle suburbs, both serviced by trains and trams, and then there are the outer-middle and outer suburbs based around private motor vehicle use. Overall, passenger mobility is now dominated by private car use for all but the shortest trips and for all trip types. Private car use is relatively high even in the inner and middle suburbs.

As described above, this form of mobility has high environmental and social costs. Reducing the environmental impacts of transport therefore entails a mix of several broad responses: overall travel demand reduction; mode switching to cleaner modes – i.e., from cars to public and active transport; and using cleaner and more efficient private vehicles. At the national and State level there is a raft of public policy initiatives designed to promote action on these three fronts, yet the overall trend of much car-based urban development continues in Melbourne. A number of commentators have described this implementation gap and sought to identify key barriers.

This problem has a number of features. Land use planning has followed a strict grouping of land uses, necessitating longer trips between residential, industrial, commercial, and service zones, as opposed to the more mixed-use development that characterizes the older, inner-city areas (Buxton and Scheurer, 2005). As described above, there are no fixed-line public transport services available in the outer suburbs. Many contemporary suburbs are designed on cul-de-sac patterns making access by public transport difficult. These suburbs frequently lack any local services, necessitating private motorized travel for all social and economic needs. Given that over one-half of the new residences have been in the outer 'growth ring' local government areas, it is where residential expansion has outstripped the government's capacity to supply services – not only public transport, but these areas have serious shortfalls in local employment, schools, medical centres, and recreational facilities. Already at a socio-economic disadvantage, the growth ring areas are falling further behind the rest of the city.

One feature of the recent decades of high population growth has been the concomitant expansion of the freeway system. Although freeways only began in Melbourne in the 1960s, even now when the wisdom of freeway expansion has been questioned, this has been the greatest era of investment in major road infrastructure. In contrast, investment in public transport infrastructure has been sparse. This is largely an artefact of the Australian federal system of governance, as the federal government has funded most of the new freeways, but the State has had to fund public transport infrastructure with the inevitable result of the neglect of the latter (Russell, 2009).[1] Road congestion appears to be politically sensitive in a city dominated by car transport and there has been much support for new roads. However, overall road travel times are increasing, suggesting that there is an induced traffic effect in operation (Low and Odgers, 2012).

Access and equity in mobility: a persistent problem

A common saying is that the measure of a civilization is how it treats its weakest members. Inequitable access to mobility is a consequence of socio-economic differences and the distribution of mobility options within the city. Before the current era of mass motorization in Melbourne, lower-socio-economic groups were concentrated in the inner and inner-middle suburbs where there was also a concentration of manufacturing employment. Within these areas, mobility was possible through active transport for shorter distances, and for those who could afford it there were buses, trams and trains. Particularly important here is the relative and close proximity of housing, employment, shopping, education and other daily services. Access to mobility in this era was often determined by affordability of public transport rather than by residential location; however, access to mobility for the handicapped was particularly poor. Mobility poverty does not feature as a major social issue in these times, although no doubt its occurrence has been under-represented and recorded.

Under mass-motorization these factors changed considerably and access to mobility becomes shaped by geographical factors. For a number of reasons, the inner-city underwent a revitalization from the 1960s onwards, with a constant escalation of the costs of housing, so that, by the contemporary era, the city's most expensive real estate generally occurs in the older, inner-city suburbs, including those that were once the preserve of the working class in the last century. Consequently, those of lower economic status who are not accommodated in public housing are now concentrated in areas of lower housing costs which are to be found at increasing distances from the city centre. An attraction of the inner city for professionals and white-collar workers was, of course, the availability of public transport. Proximity to railway stations has a positive effect on housing prices, even in the middle suburbs.

Beyond the reach of the fixed rail services there are essentially two mobility choices, public transport on buses or the private car. Active transport is often problematic given the absence of specialized infrastructure to facilitate walking and cycling, and the longer distances between residential land use and desired services when compared to the older suburbs. Bus service access is generally poorer with increasing distance from the city centre and service levels usually decrease greatly off-peak and on weekends (Loader and Stanley, 2009). A complication is that the geographical distribution of employment by type has also changed over time, such as the growth of light industry in the southeast and the new logistics-related developments associated with outer freeways. Access to inner-city jobs is facilitated by trams and trains, but access to the newer centres of employment in the newer areas of the city by public transport is often effectively impossible. It follows that there is higher car ownership in the outer suburbs and that car ownership is relatively high in lower-income households (Currie and Senbergs, 2007).

Even for the inhabitants of the new outer suburbs, the so-called 'mortgage belt', the family unit often features a couple both in employment. A notable feature of the outer suburbs is that every working member of the household has a car; the costs of car ownership and operation form a considerable portion of household expenditures. Under the prospect of peak oil and rising fuel prices, these car-dependent households face a future of increasing expenditure on mobility (Dodson and Sipe, 2007).

Perth: efforts to escape car dependency

One the world's most remote cities with a population over one million, Perth is the fourth largest city in Australia with a 2010 census population of 1.7 million. It is a fairly rapidly growing city: in the five years to 2010, the population increased by 177,000. Although positioned in a State economy dominated by resources – minerals and fossil fuels – and agriculture, Perth has essentially a service economy largely relying on inter-State or overseas imports of manufactured goods.

Overall governance of Perth resembles that of Melbourne, i.e., it is conducted by the State government – of Western Australia – which is a (bicameral) democratic parliament with a number of specialized agencies dealing with transport and land use and a government minister in charge of the main body, the Public Transport Authority (PTA). Details of these institutions are discussed below but, in overview, the PTA oversees Transperth, which operates Perth's rail system and manages the city's contracted bus and ferry services operators. Overall land use planning is largely the responsibility of the State Department of Planning advised by a statutory authority, the Western Australian Planning Commission – with some roles held by local governments and redevelopment authorities.

Perth is a prosperous city within a highly developed nation. Its urban geography, however, is characterized by suburbanization and low-density development based around private motor vehicle travel and an extensive freeway and highway system; the city stretches for some 130 kilometres north–south along the Indian Ocean. From the late 1800s to WWII, Perth's mobility was largely based around its trains, trams, and streetcars whose services served as the locus for urban development, producing a relatively small and compact city (Curtis, 2005).

Across Australia's major cities, the heyday of urban passenger rail was the inter-war period between World War I and WWII, when suburban rail carried the majority of commuting trips and was responsible for patterns of suburban development that centred on railway stations. During WW2 petrol was rationed – because it was imported; urban expansion largely stopped, and private car ownership and use remained relatively low and static. Australia's economy then boomed in the late 1940s and 50s; cities grew and the baby boom began, petrol rationing was lifted, and the Australian domestic car industry got under way, offering cheaper substitutes for imported vehicles. These transformations saw the fundamental shift in urban mobility from walking, cycling, and public transport – trains, buses, and trams – towards mass private car ownership. Urban rail ran into trouble, being plagued by the vicious cycle of falling ridership causing falling revenue, prompting service and investment reductions, which in turn further eroded ridership. In the case of Perth, these effects were exaggerated because the rapid growth in the city took hold in the 1960s, so that its urban expansion was entirely predicated on mobility by car. At the same time, absolute rail ridership was falling and its future seemed uncertain in a city of motorists.

Today, mobility is based around the private car and car ownership is particularly high. Household mobility is provided mainly by private vehicles – for example, 82 per cent of the 2006 Perth journey-to-work was by car, with only 10 per cent by all forms of public transport and less than 4 per cent by walking or cycling. Inner suburbs' public transport use for commuting approaches 20 per cent of journeys, while in the middle-outer suburbs this proportion can fall to less than 5 per cent (ABS, 2008). GHG emissions data is not available for Perth specifically, but based on the most recent Western Australian State GHG data – for 2009 – and apportioning this according to Perth's population share of the State, we can estimate Perth's transport emissions at 8.73 $MtCO_2e$ (ABS, 2011; DCCEE, 2011). On a per capita basis, this amounts to 5.3 tonnes $CO2_e$. Against the baseline of 1990, the 2009 State emissions represent an increase of 76 per cent (DCCEE, 2011).

Perth suffers from the same symptoms of having an unsustainable transport system as Australia's other capital and major cities, characterized by high environmental costs, a high degree of car-dependency, inequitable access to mobility services, and so on. One distinguishing feature of Perth is an urban form with a very high proportion of suburban dwellings and very little higher-density residential living. A pattern of southward and northward coastal urban expansion is creating a strongly linear orientation, but with the CBD still functioning as the mobility hub of this expanding and remote city. Set against the overall picture of an isolated and sprawling metropolis are several achievements in sustainable transport, notably the efforts in land use and transport planning and the expansion of the suburban rail network (discussed in Chapter 8).

Transit oriented development in Perth

Although the limitations and problems of basing urban transport on automobility have been recognized for some time, it has taken many decades for transport and land use planning to shift its focus towards the use of urban planning solutions to urban mobility. This has amounted to a mutual recognition by land use planners and transport planners of the need for integrated approaches to providing urban mobility and the rise of land use-transport integration (LUTI). At the local scale, transit oriented development (TOD) exemplifies this form of integration. Typically, TOD concentrates urban development around public transport nodes, usually in the form of residential and commercial land uses clustered around railway stations. Often the TOD precinct is defined as being within an 800-metre – or ten-minute walk – radius around the public transport station (Curtis et al., 2009). Taken up worldwide, TOD is part of the urban strategies of the major Australian cities. Implementing TOD around the world and in Australian cities, such as Melbourne and Perth (e.g., Curtis and Armstrong, 2009), has proved difficult in practice.

Against this background, Curtis (2012) examined Perth's TOD experiences up to 2007, with particular attention to describing and explaining the extent of TOD implementation. Perth is interesting for several reasons: it has strongly endorsed policy, TOD policy being well established since 1988; and three contrasting planning models have been used, allowing for comparative evaluation (Curtis, 2012). These three models were conventional planning – with implementation by private developers; development by the State; and a dedicated redevelopment authority. Given lead times of eight to ten years between TOD conception and project completion, performance assessment needs to consider activity over a longer time scale.

First and foremost, Curtis (2012) found that the walking catchments of Perth's 68 railway station precincts did not provide coverage of the majority of city precincts, so that most of the city was beyond walking distance from the nearest railway station. Most of these precincts were in low-density residential areas – with gross density defined in this study as less than ten dwellings per hectare – and of the seven catchments in higher residential density areas, only three were outside the city centre. Curtis's calculations were that 84 per cent of the precincts had residential densities less than ten dwellings/ha. Employment in the precincts was discouraging; of the 68 precincts, 13 outside central Perth contained employment uses and nearly all of these were of low work floor space density.

Curtis evaluated the extent to which State policy, and the policies of the 25 local government areas covering the railway station precincts, influenced TOD. Measured against seven TOD principles, most of the State policy documents – such as the *State Planning Strategy*, the *Metropolitan Transport Strategy*, and *Network City* – covered some of the seven criteria. But, as the author concludes, 'very few policy documents include strong statements of direction for TOD' (Curtis, 2012). At the local government level, the TOD criteria were generally not strongly supported. A general failure in these local planning schemes to support TOD is identified as a factor in poor TOD implementation.

Measures of density and intensity of land use were used to assess the extent to which the State's 2005 TOD planning targets implementation in each railway station precinct had been met. Broadly, the State's density target was 25 dwellings/ha – and higher with closer proximity to high frequency public transport. This density target was not achieved in any of the precincts (Curtis, 2012). Applying Newman's and Kenworthy's (2006) yardstick of development intensity of 10,000 employees and/or residents within a station precinct to achieve efficient public transport, the study found that only eight of the 68 stations satisfied this goal. Tellingly, all eight were in the inner suburbs and the city's CBD. A related study found that non-residential land uses at three of the new station precincts in Perth did not meet TOD goals (Curtis and Mellor, 2011). In those instances of TOD success, Curtis (2012) found institutional factors to be highly influential, an issue examined in the following section.

Institutional innovation for transport and land use coordination

As described above, there has been growing and considerable interest among transport operators and commentators in the role of integration in transport and urban systems. Integration can be identified at several levels, ranging from integration of public transport services across cities and regions – such as coordinated lines and modes, integrated ticketing, and coordinated timetables – through to techniques such as network planning. A higher form of integration is achieved by integrating active, public, and private transport modes into urban transport planning. Achieving higher order social and environmental goals appears to necessitate land use and transport integration (LUTI) so as to address such issues as offering multiple modal choice, reducing transport demand, and reducing journey lengths. Indicative of the acceptance of the LUTI principle is the existence of a National Charter on Integrated Land Use and Transport Planning. In the literature, a difference is often drawn between coordination – which can occur between stakeholders with different goals – and integration – requiring unified stakeholder goals. The latter is essential for achieving LUTI (Curtis and James, 2004).

As argued throughout this volume, institutions play a critical role in the quest for sustainable transport and their role is particularly important in transport and land use planning for LUTI. Although Perth has shared the typical Australian embrace of free market principles in its urban planning, and has been a high-growth city, it is notable for its experiment in integrated transport and land use planning (see Stone, 2009b for a contrast with Melbourne). Curtis and James (2004) provide an account of the merging of Western Australia's land use planning and transport planning agencies into a single agency – the Department for Planning and Infrastructure – under a single government minister in 2001. Its goal was to improve coordination and service delivery, yet it did not have a consolidated programme to deliver LUTI. But this experiment proved to be short-lived; the amalgamation was undone in 2009 and its legacy remains somewhat unclear. [Indeed, based on their own recent accounts of their recent activities, neither the Department of Transport nor the Department of Planning perceive LUTI as a high priority.]

Although the initiative of an integrated land use and transport department was not associated with an obvious surge in TOD success, Curtis (2012) identified the virtues of two aforementioned alternative institutional models in promoting TOD, namely the State acting as a developer through its own land development agency, LandCorp, and the use of a dedicated redevelopment agency (RDA). Two LandCorp projects in railway station precincts achieved high residential densities (Curtis, 2012) and there have been some notable RDA projects in TOD, such as the Subi Centro development in the suburb of Subiaco by the Subiaco Redevelopment Authority (Howe, Glass and Curtis, 2009).

In a city characterized by its extensive use of private cars, the transition to greater use of public and active transport in Perth through TOD has been difficult and slow despite two decades of effort. With the exception of the city centre, residential densities remain low and there is little mixed-use development or employment in the railway station precincts away from the city centre. State and local planning could play a more influential role in promoting TOD, but a greater barrier appears to be the general absence of effective institutions. Traditional State and local planning approaches have not been effective. An encouraging sign is where the State takes up the development role itself or creates an effective RDA, for under these institutions the city has been able to host successful TOD projects.

Shanghai: paradoxes in managing growth in a mega-city

Shanghai, the most populous city in the world's most populous nation, has – not for the first time in its modern history – become an economic powerhouse and exemplar of rapid economic growth and economic transformation. It is China's commercial and financial centre. With a central population of over ten million – within 600 sq kilometres – and a greater urban population of around 23 million, it qualifies as a mega-city, covering an area of 6340 sq kilometres. As a consequence of internal national migration, the city has been growing rapidly since the era of economic liberalization began in the early 1990s.

Public and active transport, combined with para-transit, has formed the basis for mobility in the modern era. Accordingly, Shanghai has one of the world's largest public transport systems. Expansion of the Shanghai Metro has occurred at such a pace that it has virtually re-written what might be deemed possible. As of 2011 (22 October), the Metro broke the world record for daily ridership with 7.548 million riders; annual ridership in 2011 was 1.9 billion. Matching the scale of investment is the scope of the city's ambitions; the metro's first 12 lines reputedly cost 93.6 billion RMB and the planned addition of a further 17 lines – covering 780 kilometres – is expected to cost 225.5 billion RMB (de Jong et al., 2010).

Mass motorization, despite being at levels below that of developed Western cities, has reached the stage where it is a blight on many aspects of Shanghai daily life. Private car ownership is booming as individual prosperity increases, yet road congestion, collisions, and air pollution are rife – albeit, less so than in Beijing. Despite the expansion of mass public transport, many of the new town districts have poor public transport access. Furthermore, the city has a considerable stake in the motor industry. Shanghai has a major motor industry in its own district, the Shanghai International Automotive City – which also contains China's Formula One car-racing circuit. In these ways, Shanghai is as conflicted over the growth of the private car as many of its Western counterparts.

Han (2000) describes the complex arrangements behind Shanghai's urban development, arguing that it is neither a model of centralized planning nor an exemplar of free markets, but a product of both. Although it is subject to national policy directives and receives national financial support, transport in Shanghai is largely a matter for the city's own government. Essentially, the primary institutions governing the public transport system are the public agencies of Shanghai, as would be expected in the Chinese economy. A Bureau of Transport is responsible for management of construction and is the overall administrator of the metro, bus, and ferry companies. Shanghai Metro is operated by two companies: Shanghai Metro Operation Company (Lines 1, 2, 3, and 4) and Shanghai Modern Rail Transit Co. (Line 5). There is a Bureau of Municipal Engineering that deals with oversight of construction and maintenance of the system infrastructure – except for the metro system. Security services and maintenance for some facilities are provided by a Bureau of Public Safety. Coordination tasks are handled by the Shanghai Urban Transit Bureau.

By virtue of the city's scale, and despite the powers that its government can exert over public and corporate activity, the sustainable transport challenges facing Shanghai are simply enormous. Although popular attention has dwelt on the growth of private motorcar use and acknowledging that this is a serious issue, moving to sustainable mobility is a wider and more complex problem. Modal splits within commuting also tell us something of a city where the old and new co-mingle; 36 per cent of commuters use the slow modes of walking, cycling, and electric bicycling, while 20 per cent now travel by motorcar. Much of this problem is attributed to the high rate of economic – and population – growth, but more fundamentally what is occurring is an economic transformation accompanied by a re-structuring of the urban economy. Yet, Shanghai's response to these issues is equally impressive.

Twenty-first century metro

Speculations that the twenty-first century might be dominated by the story of China can draw on few more potent demonstrations of the nation's aspirations and capacities than its achievements in building the new metro systems in its major cities. This nationwide effort is the largest national investment in public transport ever undertaken. And there is no better example than the case of the Shanghai Metro for the sheer speed and scale of what this city undertook. Shanghai's first metro line opened in 1995 and the Shanghai Metro is now the world's largest network in terms of passenger route length (Although Tokyo's combined metro and subway system still exceeds Shanghai's metro).

Investments in metros are an important part of the nation's economic development strategy for this rapidly urbanizing population. Although Shanghai is home to motor vehicle manufacturing, there is a deliberate strategy to provide mass mobility based on an extensive subway system. In some ways, Shanghai is drawing on Western experiences with providing urban mobility – but those of the nineteenth century rather than those of the latter twentieth century; i.e., the use of underground railways – subway systems – to accommodate rising mobility demands in the major cities before the advent of extensive car ownership, such as in London, New York, Moscow, and Paris. In a historical reversal of sorts, very few major cities in Europe or North America with existing subway systems have invested in their expansion, with London, Madrid, Barcelona, and Paris being the exceptions.

What has enabled Shanghai and other major Chinese cities to build entirely new metro systems with such speed has been in part due to the ways in which these projects are financed and managed. Finance has drawn on national, provincial, and local funds – as well as foreign funds in Shanghai's case – and the projects are managed by State-owned enterprises (SOEs), that also have the capacity to borrow from banks. Being State-guaranteed, the SOEs enjoy a number of advantages, including access to low interest rates, the ability to endure project cost overruns, and the accumulation of high debt levels (de Jong et al., 2010). Although, as de Jong et al. note, there are rising criticisms of the SOE model and questions as to whether they will be able to continue to operate the same way into the future.

Ridership is high and, as is typical of most Chinese metros, the fares are low. However, the Shanghai system is not ideal by any means. Many stations are without platform screen doors, and facilities for disabled access vary considerably across the system. Further, there may be some planning problems emerging, such as line locations mismatching with places of higher population growth, orbital lines not being as effective as they ought to be, some lines being ill-suited to high density ridership, and some alleged instances of overcrowding and inefficient transfers between lines. Also because of the mismatch between service locations and urban development, there are fewer passengers towards the extremity of the system, unbalancing passenger flows, and there is considerable short-distance metro travel.

There has been considerable discussion of the capacity of cities such as Shanghai for such rapid construction. One response is that such achievements are a function of a one-party State and the advantages such a political system offers. These advantages include the ability to direct large amounts of public funding to strategic public projects, the effectiveness of political processes that are without opposition, and the overwhelming powers of the State in any disputes with private parties. All these advantages are in evidence in Shanghai. Others point to the low labour costs in China as a major factor. But as a counter-argument, there have been other cities and political systems also capable of high investment in subways and mass public transport in the post-war era, such as in Tokyo and more recently, in Spain. Further, as supporters of public transport have pointed out, there is an inconsistency in claiming that public transport investment only flourishes in authoritarian States, while at the same time allowing that major investment in public road infrastructure readily occurs under democratic governance.

In short, it is difficult to generalize about China's unique system of governance, having as it does features of single-party State authoritarianism, State-sponsored and State-directed capitalist development, and centralized party control over major appointments to major public entities. Despite the power of the Communist Party's Standing Committee and the People's Congress at the national level, local governments have considerable autonomy and city authorities have demonstrated remarkable organizational skill in handling these transport mega-projects.

Mass car ownership in a centrally planned economy?

Many commentators have been concerned with the environmental implications of rising private motor vehicle use in rapidly industrializing nations. Increasing private wealth formation resulting from economic growth in India and China has drawn particular attention as private motorcars are being acquired for private mobility and as symbols of material success (Pan, 2011). Until the 1990s, car ownership in developing nations typically implied imported vehicles owned by the wealthiest or most highly privileged. But this picture has changed quickly. Private cars are no longer the exclusive preserve of the wealthy as increasing domestic manufacturing in developing nations is giving rise to low-cost local vehicles, combined with the availability of low-cost vehicles produced by established manufacturers – such as Fiat, Hyundai, Renault and Suzuki. India's Tato Nano retails for the equivalent of a few thousand US dollars, for example. China is now the world's largest automobile manufacturer – producing 13 million units annually – with around ten China-based manufacturers and another dozen or so joint ventures with international firms. It is the goal of these manufacturers to make private motoring available to those who were never previously able to afford such assets. In addition to urban passenger transport, mass motorization is also occurring through small enterprises that are taking up the use of motor vehicles, especially light commercial vehicles, to replace two-wheelers or non-motorized vehicles.

Ownership and use of private cars in Shanghai has risen rapidly from a relatively low base by western standards. From 1990 to 2000, the city increased road length by 40 per cent but the total number of motor vehicles increased 2.5-fold – to over 700,000 vehicles (CFPTVC, 2003: 141). At that time most motor vehicles were owned by governments and businesses; only 15,000–50,000 motor vehicles were privately owned (CFPTVC, 2003: 231), but we should note that this wide range accounts for some 30,000 vehicles registered to motorcar companies but owned by their employees. In the early 2000s, Shanghai had low per capita private vehicle ownership due to the expense and difficulty of obtaining a driving licence, the relatively high costs of ownership and use – in a country where consumer credit was not yet widely available – and the levels of road congestion.

But several things changed to lower these barriers. National economic policy to develop car manufacturing capacity led to policies to encourage domestic ownership; and Shanghai hosts several manufacturers. Available credit, lower imported vehicle costs, the high status attached to vehicle ownership in a city economy of growing wealth, and the city's 'multi-centralization' policy and more dispersed population all encourage private ownership and use. As a result, Shanghai's vehicle population grew by almost one million vehicles in the decade to 2008 – from 0.49 to 1.47 million (Hao et al., 2011). Carbon dioxide emissions from passenger vehicles grew 2.6 times between 2000 and 2009 (Wang et al., 2011).

Shanghai's problems arising from substantial private motor vehicle use are not dissimilar to most large developed-nation cities: congestion of urban space, air pollution, high greenhouse gas emissions, declining public transport share, large public investment in road infrastructure, and decentralization of housing and employment. Urban space in Chinese cities planned before the late 1980s had high residential densities and comparatively little space dedicated to streets, as the CFPTVC (2003) points out, with little expectation of accommodating large numbers of motor vehicles. Road congestion in the older parts of the city can therefore occur quite readily as traffic volumes increase and there are increasing conflicts over the uses of these streets. A distinguishing feature and one under-appreciated in the West is that the rate of motorization itself is a source of problems. Developed nations have had many decades to adjust to mass motorization through planning and investment, whereas cities such as Shanghai are experiencing comparatively greater problems of mass private vehicle use at far lower levels of car ownership than in developed nations (CFPTVC, 2003).

One aspect of the city's responses has been the familiar – and largely discredited – approach of increasing road infrastructure capacity. Consequently, the inner parts of the city are now dissected by elevated freeways. Policies restricting road use to motor vehicles have also been used to improve vehicle throughput by banning bicycles from major roads. Growing demand invariably undermines such strategies, accelerated by the 'generated traffic' and 'induced demand' effects. More promising are demand management strategies, and Shanghai has turned to some of these.

Government officials moved to address motor vehicle traffic comparatively early relative to vehicle numbers. Shanghai's first efforts to curb private vehicle use began almost two decades ago in 1994 using a licence plate quota system for private cars allocated through auctioning – although some sources claim the introduction occurred in the mid-1980s. The monthly plate auctions typically offer between 5,000 and 8,000 new plates, with winning bids ranging from 23,000 RMB to 56,000 RMB (around US$3,600–8,500). Combined with rush-hour controls on non-Shanghai registered vehicles, this measure effectively serves as a congestion charge. As the city's economy and the wealth of its citizenry have grown, these policies have been subject to considerable pressure. Also, the system has been subject to abuses, such as hoarding by car dealers with resultant profiteering. Locals have also found ways to circumvent these restrictions, such as by licensing their cars in locales outside Shanghai. These schemes were replaced by limits on new car and truck registrations and special measures to curb motorcycles and scooters.

Such small engine capacity mopeds and scooters have boomed across Asia, especially in places where bicycles were popular, providing cheap personal transport. They are also used by many small enterprises for light goods freight transport. Often powered by two-stroke engines, the cumulative effect of the emissions of large motorcycle and scooter fleets has been very detrimental to urban air quality. Shanghai has banned petrol-powered bikes since 2006, with this market now served by electric bicycles ('e-bikes') or gas-powered bikes. There were several responses by the city authority including capping registrations, effectively limiting ownership of vehicles, and banning the small capacity two-wheelers from the city centre; exceptions have been made for those that are battery-powered.

Clearly, the future of the private car in Shanghai is a story in flux. It is noteworthy that Hao et al. (2011) found that fuel consumption in Shanghai will be reduced by 59 per cent by 2020 compared to a 'business as usual' case without policy intervention. However, it is the opinion of some transport experts that Shanghai's policies have been effective in reducing car ownership by comparing the city with Beijing where car ownership is several times higher. We need also to remember that the rise of the motor vehicle isn't exclusively an issue with private cars, as fully one-half of CO_2 emissions from the public transport sector are from taxis.

Singapore: innovation and the Singapore model

All the economic and social development stories of Southeast Asia's tiger economies are remarkable in their own ways, but that of the city-State of Singapore has drawn much attention, giving rise to the 'Singapore model' that combines centralized government planning and free market capitalism. Comprising one major and many small islands, Singapore covers about 714 sq kilometres – about 40 kilometres east–west by 25 kilometres north–south) and had a 2012 population of 5.2 million. Sustained high rates of growth have created a prosperous and dynamic economy from the newly formed developing nation with few natural resources. From the formation of the Republic in 1965 up to the present, Singapore became one of the world's richest nations on a per capita GDP basis. Although the Singapore model concerns economic goals, it has also exerted a strong influence on the island's land use and transport planning and has allowed these to contribute to Singapore's prosperity and reputation for high urban amenity. Indeed, in what is effectively a single party and authoritarian State, some have suggested that the social contract between citizen and State is founded on expectations of household prosperity and a high quality of life.

There has only been one Party in government in Singapore's Westminster (unicameral) parliamentary system since the nation's independence: the People's Action Party led by three successive prime ministers since 1959. Overall governance of transport falls under the Ministry of Transport, with a statutory board handling all land transport – the Land Transport Authority – and another statutory authority, the Public Transport Council, covering the regulation of public transport – including fares and licensing. Public transport operations are divided between SBS Transit and Singapore Mass Rapid Transit Trains, and Trans-Island Bus Services.

Singapore is often referred to as a special case in Asia because of its unusual geo-political status as a city-State on a small island and its centralized political system. However, the strongly interventionist approach of the State is not uncommon across Asian countries. Heavily centralized government, successful public direction of economic growth, the long reach of government authority and a highly concentrated urbanized population have created opportunities for public sector innovation in transport and land use planning. Singapore-style initiatives have in fact been adopted by Western nations; London and Stockholm referred to the Singapore system when introducing road congestion charges (Barter, 2008). What *is* distinctive about the Singapore model is that it aims to manage both public and private modes of transportation rather than necessarily prioritising one over the other.

Singapore's land use planning and transport systems are among the most studied in the world. Under its centralized State planning and regulatory system, it has been both a careful investor and willing innovator in using both market and regulatory instruments. Its urban development has been carefully controlled and is concentrated in high-rise satellite towns built around public transport hubs. Its high-capacity rail system is integrated with an extensive bus system. There has also been investment in high-quality road infrastructure. There are 3,200 kilometres of paved roads and more than 700,000 motor vehicles (NEA, 2010). A number of policy instruments have been skilfully applied to limit private ownership of cars and restrict road usage in selected areas of the central city.

Travel behaviour has been changing over time and not necessarily always towards greater sustainability. Over the period 1997–2008, total daily journeys increased greatly, by 32 per cent, and outstripping the 26 per cent population gain (Choi and Toh, 2010). Public transport's daily mode share dropped from 63 per cent to 56 per cent, which is partly explained by the growth in the car population: although stable between 1998 and 2003, between 2004 and 2008 car use increased 32 per cent (Choi and Toh, 2010). As described below, these developments prompted government initiatives to curb car ownership. Within public transport there was also a shift away from buses and taxis to rail: in 1997, on a per capita basis, buses had a 61 per cent share which fell to 55 per cent by 2008; the shortfall was taken up by the rail system (Choi and Toh, 2010).

According to its 2008 *National Climate Change Strategy*, Singapore's GHG emissions from its transport sector in 2005 were 7,056,000 tonnes CO_2 in combusted fuel and another 930,000 tonnes derived from electricity use, totalling 7,986,000 tonnes (Singapore, 2008). The transport sector accounted for some 19 per cent of national emissions (ibid.). Han (2009) contends that Singapore's model of coordinated policies and the mix of public and private modes offers a model for sustainable transport governance in Asia. In terms of sustainable transport performance, however, there are grounds for concern over the extent of freeway investment, a considerable private motor vehicle fleet, and the long distances covered by private vehicles each year, as well as relatively low use of walking and cycling, prompting the government to respond with various measures (e.g., Choi and Toh, 2010; LTA, 2008; Singapore, 2008, 2010).

Excellence in public transport

Singapore is renowned for its high-quality public transport, which forms the backbone of passenger mobility across the city. Although most investment has gone into the Mass Rapid Transit (MRT), there is also a small light rail system (Light Rapid Transit,or LRT), and an extensive bus system and a large number of taxis running on public roads. In 2010, there were 3.8 million daily public transport passenger journeys at an average of 9.8 kilometres travelled per journey (LTA, 2011). The breakdown in terms of mode from the same document provides the following figures in passenger trips: 3.199 million by bus, 2.069 million by MRT, 0.912 million by taxi, and 0.1million by LRT (LTA, 2011).[2] There are slightly fewer daily train trips per capita than bus trips – 0.4 and 0.6 respectively – but these cover greater distances – 10.3km versus 4.8km.

Although the Singapore model highlights government efficiency, it is interesting to note that there was much deliberation over plans for mass transit that began with a 1967 study on rail transit. Feasibility studies took place from 1972 to 1980 on a rail and bus system, but a contrary review advocating a comprehensive bus system prompted further study. By 1982, the government had settled on the rail and bus system and allocated S$5 billion for a 67-kilometre MRT system. The MRT opened in 1987. Work continued on the public transport system and today there are 130 kilometres of MRT and 29 kilometres of LRT (LTA, 2011).

Together, the MRT and LRT networks form the Rapid Transit System (RST). The Land Transport Authority plans to double the RST network to 278 kilometres by 2020 through adding new lines and extensions (LTA, 2008). A complementary policy is to increase frequencies of train services within the limits of the operator's investment to reduce crowding and increase passenger comfort. The cornerstone of the system is the MRT. Operational since 1987, it currently has four lines working. Buses feed services to the MRT, with taxis providing 'high-end' services. The LRT is smaller in scale and spatial coverage. Its purpose is to increase the catchment of MRT railway stations by providing the feeder services.

The Public Transport Council (PTC) sets a standard to guide the two main bus operators. For example, towns with predominantly public housing – Housing Development Board projects – must have MRT and/or bus services that directly connect with the city centre, bus stops should as far as practicable be 400m apart and feeder bus services must meet the last train of the evening. The MRT network was built after a 10-year debate between consulting teams over 'buses versus trains'. The main issue was the cost-benefit outcomes that might be expected from the MRT (Richmond, 2008). Sophisticated and expensive technology is a key part of prioritizing public transport in traffic flows. 'Smart' traffic lights detect approaching buses, and turn the lights to green. A 'B' signal comes on before the green light, giving buses a head start on other traffic.

Integrating land use and transport

From the date of independence from Malaysia in 1965 to 1971, the government focused on housing and employment rather than transport and was without an overall plan to integrate them all. During this period, however, the basis of the urban form of present-day Singapore was laid out. The location of residential and industrial construction reflected a clear intention to break from the past by extending development island-wide, rather than be confined to a central area as planned by the last colonial government in 1958. In 1971, the first concept plan was formulated with advice from the United Nations Development Program (Han, 2005). This plan was the first blueprint to guide land use and transport planning. Its main features included the organization of land use into high-density groups that were connected through the transportation network. Future rail extensions would be integrated with land use outcomes as the rail-based MRT linked new town centres, while the LRT increased the catchment area of MRT stations.

A Concept Plan is used to direct long-term planning with a 40–50-year planning horizon for land use, covering such themes as transport, housing, infrastructure, and recreation; the last was released in 2001 and the next, the 2011 plan, is still under development at the time of writing in 2012. Detailed land use planning is based on the 2008 Master Plan administered by the Urban Redevelopment Authority.

Taming the car

Singapore, as a highly urbanized small island with a prosperous economy, is clearly at high risk from the problems of high levels of car use. Historically, rising national incomes tend to be associated with increased car ownership (Kutzbach, 2010). Recognizing this risk, Singapore's government responded with a strategy of constraining car ownership and controlling road use. As a consequence, Singapore does not suffer from the same problems of road congestion that plague many of the region's major cities, such as the extreme cases of Bangkok, Beijing, Jakarta, and Manila. There were 110 vehicles per 1000 citizens in 2011 according to the national Department of Statistics, which is about one-fifth of the OECD average.

Car ownership is restricted through government policy and a number of specific taxes make private cars expensive to own. In the early 1970s the government used increasing levels of purchase and ownership taxes to curb ownership. Then, in 1990, Singapore introduced the Vehicle Quota System (VQS) that controls the absolute numbers of vehicles able to use the nation's roads; it was the first of its kind in the world. Using a competitive bidding system, the potential buyers bid for a Certificate of Entitlement that allows ownership of private vehicles for ten years. After the certificate expires, the vehicle must be de-registered or a new certificate purchased. There are categories of vehicles according to engine size and vehicle use, although scheduled buses, school buses, and emergency vehicles are exempted. As an indication of certificate costs, certificates for large cars – 1600cc engine capacity and above – were priced at around US$52,000 in early 2012. The universal coverage of the certificates makes it impossible to evade the strictures of the scheme. Although initially transferable, resultant speculation and profiteering prompted officials to make the certificates non-transferable. By exercising control over the number of certificates, the vehicle population and its rate of growth is controlled. Additionally, Singapore's car owners are subject to other fees and taxes: an import duty of 41 per cent, annual registration fees, an additional registration fee for the first registration, and an annual road tax. Before 1990, the number of vehicles had been increasing at 7 per cent annually; for the three years after 1990, the rate was 3 per cent (Sharp, 2005). Growth in vehicle numbers has exceeded population growth and has been considerable. In 1990, there were 708,000 motor vehicles of all classes – including 399,000 motorcars – which by 2011 had grown by 35 per cent to 957,000 – with 592,000 motorcars. Olszewski (2007) showed that per capita energy use in the transport sector was effectively curtailed during the first seven years of the VQS.

Road access is also controlled through road pricing, which effectively began in 1975 with the Area Licensing System covering 725 ha (Han, 2009). This manual system used pre-paid, displayed tickets and focused on CBD access. Encouraged by early success in reducing congestion, the authorities extended the scheme to other major congested roads with the Road Pricing Scheme. When multi-lane electronic monitoring technology became available this was installed and the Electronic Road Pricing programme was initiated in 1998 on the major roads entering and exiting the CBD. Vehicles have an in-vehicle unit and sensors are installed in overhead gantries; charging is handled by smart card technology. Average daily traffic flow declined by around 20 per cent in the first month of its operation in 1998 (Foo, 2000). Singapore's major arterial roads and freeways are relatively congestion free in peak periods. Ongoing experimentation has demonstrated that traffic flows can be manipulated by changes to the charging regime (Han, 2009).

Controlling motorization, however, is clearly not the same as minimizing car use. Private cars may be restricted and a luxury item, but the average Singapore car covers 19,100 kilometres annually (LTA, 2011). Given Singapore's relatively small area, this usage level implies a considerable number of journeys. Driving is no doubt popular in part because the major roads are typically free of congestion and the nation has invested in a steady expansion of high-quality roadways. New highways and increased road network capacity were constructed steadily as the nation as a whole developed. The 3297 kilometres road network covers some 12 per cent of the island's land area (Han, 2009). There are 161 kilometres of expressways and another 634 kilometres of arterial roads (LTA, 2011). According to the Land Transport Authority, in the 20 years from 1990 to 2010 the length of freeways increased from 104 to 161 kilometres, arterial roads increased from 529 to 634 kilometres and collector roads grew from 260 to 535 kilometres (LTA, 2011). Between 1986 and 1996, the road surface area increased by 27 per cent and S$3 billion was invested between 1996 and 2000 to construct a further 300 lane-kilometres of roads (Han, 2009). The government plans to scale down the road expansion programme from the current 1 per cent to 0.5 per cent per annum over the next 15 years and reduce the vehicle growth rate from 3 per cent to 1.5 per cent after 2009 (Lew and Ang, 2010).

Discussion and conclusions

We began this chapter by noting the differences between our case study cities. As shown, these four cities demonstrate a variety of mobility experiences, institutional/ political practices, and public policy responses to transform their transport systems. Mobility and land use in these cities is indelibly the consequence of strongly local influences of culture, geography, and history.

Despite these differences, what is also inescapable is a certain commonality of social, economic, and environmental challenges. In some ways this is unsurprising, given that all these cities are experiencing the costs of ongoing modernization – i.e., *motorization* – of their urban mobility systems, driven by both domestic economic development and by a close coupling with the global economy and the forces of globalization. And, not unusually for cities of the Asia-Pacific, the modern history of each city owes something to the post-colonial experience; indeed, all spent some time under British tutelage. All these cities can be considered global cities in the context of their close ties to the global economy. Ironically, although Perth is relatively small and a quite geographically remote settlement, it is arguably the most dependent on international economic inflows of capital and the benefits of Western Australia's mineral and energy exports.

Further linking these cities are problems of growing private motorization, at least in gross measures of numbers of registered private motor vehicles; increasing GHG emissions from transport sources; a dependence on fossil fuel energy for road transport; economies that are highly dependent on international freight transport – notably shipping in all cases; and high and growing demands for residential land uses. This is not to claim that these cities are on the same journey, but their mobility challenges are not as diverse as might be first assumed by a casual observer or those only examining isolated mobility issues.

As issues of governance, these case study cities are instructive, given the range of experience. Broadly, the issues covered here relate to the problems of creating integrated public transport systems, of finding ways to tackle the existing or rising use of private cars, and the challenges of developing urban forms to facilitate low-carbon transport.

These cities have demonstrated the tension between endogenous conditions of culture, history, and public institutions – and the problems created and solved within the local context – and the exogenous factors impinging on cities created by global environmental problems, global political institutions, international economic ties, technological innovation and international diffusion, and the influence of communication and information exchange. Divisions between local and global influences have become less clear in contemporary times.

Held against some measures of environmental performance, such as success in reducing greenhouse gas emissions from the transport sector, none of these four cities has achieved exemplary outcomes, as their emissions continue to grow. On some social indicators of sustainability, such as providing public transport access to a wide proportion of the population, there are clear examples of success. The major cities of the Asia-Pacific, as shown by our case study cities, are all struggling to come to terms with the major challenges of environmental sustainability and have adopted a range of responses.

It is the variety of responses and solutions that are instructive. Mixed with successes and innovations described here are the lessons of the other kind – of mis-steps, lost opportunities, and failures. To court a cliché, it is clear that cities are laboratories of invention and in this review we can grasp the span of these innovations and learn something of what contributed to their success. Lessons can be drawn from different cities and applied elsewhere. One lesson the cities certainly teach us is that purposeful, well-informed and planned intervention by governments is necessary to deal effectively with the problems of urban transport. Such interventions need not be associated with authoritarianism. How such decisive policy planning can be linked to democracy is explored in the next chapter.

Notes

1 There have recently been promises of federal funding for urban public transport infrastructure, notably new railway lines from West to East Melbourne. The actual expenditure has yet to result in construction as of 2012.

2 A footnote to the Singapore Table says, 'A journey refers to a series of consecutive trips on different modes (MRT, LRT and bus) made by a commuter from his origin to his destination'.

8 Action strategies for paradigm change

John Stone and Crystal Legacy

Introduction

In this chapter, we explore the means by which some urban regions have been able to make a break from the typical trajectory of auto-dependence. The difficulties inherent in making such a move are clear: in only a handful of locations can one find good evidence of conscious change in urban transport policy that has achieved firm progress towards sustainable transport. However, such success stories do exist and case studies of both 'success' and 'failure' in urban transport planning can be used to identify some 'action strategies' for paradigm change. And, to allow generalizations from these case studies, modern theoretical insights into the role of local politics and the institutional context of public policy help to build a bridge between the understandings of the systemic forces for and against change that were described in Chapter 5 and the experiences of local actors.

The case studies include six exemplar cities in northern Europe, for which we rely on analysis by the German researcher, Stephan Bratzel (1999), and three cities in the Pacific region: Vancouver, arguably North America's most promising 'sustainable transport' city; and Melbourne and Perth, two Australian cities where the politics of transport reform have produced very different outcomes. Evidence for these cities is drawn from work of GAMUT researchers (Legacy, 2010a, b; Stone, 2008, 2009b).

The examination of the process of policy formation and implementation in this broad range of cities leads to some interesting insights. Common to all success is consistent application of policies to support alternatives to car-dependence over periods of a decade or more. Other influences such as wealth or cultural attitudes to mobility appear to be less important determinants of change – or of failure to change. The key to this process is the development of particular programmes and institutions that are appropriate to the local context, and which, through their emerging success, extend the constituency of support for the new transport paradigm.

The process of change in urban transport policy is far from an orderly advance in the face of new policy learning or emerging environmental, social and economic problems. Instead, it is best characterized as an almost constant process of *challenge and resistance* between the supporters of mutually exclusive policy packages based on fundamentally different approaches to public and private investment in urban development. To understand the ground in which this challenge and resistance takes place, it is useful to use a conceptual model that encompasses such factors as the membership and structure of local political power relations, historical patterns of transport investment and urban development, and the political and institutional responses to any crisis in the legitimacy of existing urban transport policy. Such a model can be used to provide powerful explanations of past trajectories of conflict over urban transport policy in an urban region (Stone 2008).

A theoretical base for this model can be built up from a number of strands of political science and public policy analysis. This theory, briefly reviewed in this chapter, helps us to understand the variation, even among cities which in many regards might be seen as similar, in the pace and extent of transport policy change. It reinforces our understanding that considerable political and social momentum is required to hasten the process of change from the dominant conceptions of transport policy that are held by many government and professional agencies. Useful analytic tools include the regime theorists' broad conceptualization of contemporary urban governance (Leo, 1998; Stone, 1989) and the models of the complex dynamics of institutional behaviour put forward by modern analysts of the public policy process (March and Olsen, 1996; Sabatier, 1999). From this perspective, policy change is not simply a question of demonstrating the rationality of the alternative – although this is important work in building local political support. Rather, mutually exclusive conceptions of transport policy are contested in the local political space, and the potential for successful challenge to the tight institutional control typically exerted by established policy networks over problem definition and 'solution sets' appears to exist only during infrequent local 'windows of opportunity'. Once a particular policy position – typically the automobility paradigm – gains the ascendancy, a challenge group faces a very difficult task.

Both theory and practical experience demonstrate that the fundamental determinants of successful action to change the direction of urban transport policy are the choices in strategy and tactics made by political entrepreneurs to create and exploit these opportunities.

Some attempts at change: success and failure

The following case studies are all prosperous urban regions in the developed world: six in Europe, one in North America, and two in Australia. All have had access to mass automobility since the mid-twentieth century. All have seen strong citizens' movements in support of changes to policies of automobility. What varies most is the extent to which the ambitions of these citizens' movements have been translated into changes in transport policy and transport behaviour. For each, we provide some indicators of long-term changes in transport behaviour, and describe the political and institutional path of change – or resistance to change – in transport policy and its implementation.

Indications of success or failure

Direct measures of success, such as specific information about a reduction of greenhouse gas emissions from urban transport, are not readily available. Even, for example, a measure such as change in per capita vehicle kilometres travelled (VKT) is seldom reported. Therefore, in this analysis, we use a less rigorous, but more consistently reported measure of success: namely, a mode shift away from the car towards more sustainable modes.

Theoretical frameworks in urban planning allow us to use a further surrogate measure of 'success'. Following Vuchic (1999), success can be defined by the extent to which an urban region has pursued a coherent package of incentives for transit and disincentives for car travel within a set of complementary policies for both land use and broader economic and social goals. So, on these measures, how well have the cities in our case studies performed?

Six European urban regions – Amsterdam and Groningen in Holland, Freiburg and Karlsruhe in Germany, and Zürich and Basel in Switzerland – showed considerable success in reducing traffic volumes and increasing the use of sustainable modes in their inner cores in the two decades from the early 1970s (see Images 8.1 to 8.7 in the Companion Website). And, although traffic volumes grew across the whole agglomerations surrounding these cities, the rate of growth was much slower than other comparable cities (Bratzel, 1999). Zürich is the exemplar in this field (Mees, 2010, Chapter 8). All these cities show a strong degree of integration between the objectives of transport policy and of planning for urban development. All have created strong institutional mechanisms for cooperative planning and the delivery of consistent funding for transit operations and infrastructure – exemplified by the emergence of the *Verkehrsverbund* (transport alliance) model (Buehler and Pucher, 2011; Pucher and Kurth, 1996; Stone, 2011).

Vancouver has some apparent obstacles to building transit use including a relatively small CBD employment market and limited rail infrastructure. However, even in the 1970s, operators were able to reverse a decline in transit tripmaking (see Figure 8.1). Transit use peaked again in the late 1980s and stagnated in the 1990s. A trend of modest but steady growth has continued since 2001. Land use and transport strategies in Vancouver match Vuchic's prescription for success. Incentives for transit have been achieved through the maintenance of a strong urban boundary, some concentration of development in activity centres (GVRD, 1996), and continuing investment in locally proven transit technology and service patterns (GVTA, 2004). The principal disincentive for car travel is the city's controversial decision not to build commuter freeways – there is only one (the long-established Trans-Canada Highway) which traverses the city but comes no closer than 5 kilometres to the CBD – and the city's explicit commitment to using congestion as a travel-demand management tool (GVRD, 1993). Many specific targets for transit use, set in the

1990s, are now being met (City of Vancouver, 2006b). Trends in travel times for the journey to work are better than other cities in Canada, and, although data are unavailable, are very likely to be much better than any city in North America or Australia. The average time for the round trip to and from work fell from 70 minutes to 67 minutes between 1992 and 2005 (Turcotte, 2005). More recent data show that journey times have lengthened (Turcotte, 2011), perhaps due to the greater use of transit, but the long-term trend still represents a considerable achievement and shows that Vancouver has made more progress than most cities towards the 'holy grail' of self-containment. This, along with continued investment in rapid transit, gives cause for optimism that the city can in time meet the challenges of climate change and peak oil.

Since 1976, Melbourne has built more lane-kilometres of urban freeway than any other Australian city: a dramatic 'car incentive'. Despite the benefits apparently offered by Melbourne's extensive train and tram systems, annual per capita tripmaking declined rapidly after 1950 and stagnated for 25 years from 1980 (see Figure 8.1). Public transport patronage has grown since 2006 – reaching 116 trips per capita in 2010. However, this is due largely to rising petrol prices, community response to the climate crisis, and other factors extraneous to the planning and management of the transit system (Gaymer and Kinnear, 2009; Stone, 2010; Stone and Mees, 2011). Service improvements lag well behind demand. For the journey to work, census data since 1976 shows that Melbourne experienced the largest proportional decline of any Australian city in public transport use and the biggest increase in driving. The transit mode share for work trips fell from 24.1 per cent in 1976 to 13.9 per cent in 2006 (Mees et al., 2008). Melbourne's public transport problem is the inefficient management of its enviable – heavy rail and tram – resources, particularly through fragmented intermodal service delivery, which creates significant 'transit disincentives' (Lazanas and Stone, 2010; Mees, 2010, 2000). More large freeways opened in 2008, others are proposed; and an urban growth boundary established in 2003 has been abandoned. Melbourne clearly demonstrates the reverse of Vuchic's ideal.

Perth remains a car-dominated city, with a higher level of driving for work trips than Melbourne, but there are some positive signs. Per capita transit use in Perth grew by 59 per cent between 1992 and 2010 following significant new investment and strengthening of the transit-management culture (see Figure 8.1). Perth policy makers were firmly on the 'car incentive' trajectory until the 1980s. Since then, there has been a steady implementation of 'transit incentive' policies; primarily through rail electrification, construction of new rail lines, and reconfiguration of bus services (Stone, 2009b). While not wanting to overstate the degree of change, there have been some moves to slow the city's substantial road construction programme, and planning processes are tentatively engaging with the difficult task of building 'car disincentives' into a city that already has a virtually complete freeway network.

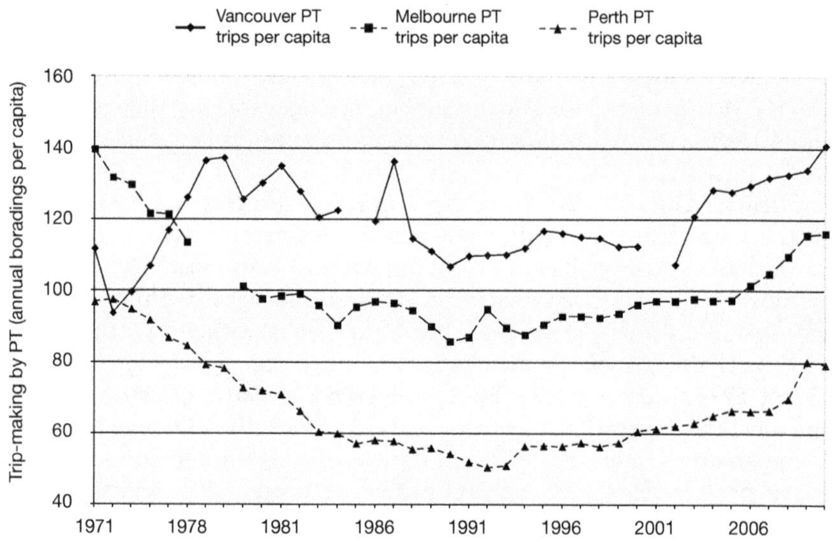

Figure 8.1 Per capita public transport tripmaking in Vancouver, Melbourne and Perth (Data obtained from Annual Reports of operators)

The contest over transport policy: political and institutional change – or not!

Bratzel (1999) identified a common pattern in the political process in all six of his 'successful' cities as each city made the shift from an automobility paradigm towards the implementation of programmes that have led to growth in the use of sustainable modes. He observed three stages of political activity and concluded that all were necessary for significant changes in travel behaviour to occur. These stages were:

1 Well-organized popular opposition to existing car-based transport policies.
2 A radical change in political leadership and a change of incumbency in important institutional positions relating to transport policy.
3 Institutionalization of new operational practice – this required the new political leaders to maintain their mandate for new transport policies while the new agenda was established.

In Melbourne, considerable effort was required in the post-war years to create the institutional capacity to construct new roads for the expanding car fleet, and against that an equally strong, but ultimately unsuccessful, social movement in opposition to the freeway plans (Davison with Yelland, 2004). As car use grew through the 1970s, deficits in the operation of urban transit also grew, and the government, with support from road agencies and the media, proposed severe cuts to services. Strong civic opposition, cleverly orchestrated by the public transport unions, forced the government to delay these cuts before an election in 1982, which saw a reformist government elected after 27 years of conservative rule. Public transport was high on the government's agenda, but the new Minister was unable to deal with the immense task of overturning many decades of 'perverted competition' – in Davison's words – between the two great statutory authorities that operated the city's large train and tram networks, and of shifting the focus of public transport towards a more productive rivalry with the car (Davison, 2004: 198). He and his successors became mired in managing the union and community backlash against measures designed to cut the costs of public transport operations, and the government was unable to maintain public confidence in its management of the transit system.

In the other main issue in the transport portfolio, the Minister, like most of the people from whom he took advice, supported the continuation of the freeway-building programme and so was at odds with the policy platform his party had taken to the 1982 election. Only a few weeks after taking office, he told an amazed group of road engineers to ignore the Australian Labor Party policy and continue work on their current freeway project (Stone, 2009b: 426).

Without support from political leaders, the various civic action groups could not mount a new challenge either to the established transit managers, who offered little leadership on alternative visions for the city, or to the roads policy community that enjoyed clear, if covert, political support and established direct influence over transport and planning policy. Over time, political leaders and the public lost confidence in the ability of managers to rebuild the transit system, opening the way for a largely unsuccessful neoliberal experiment in 'franchising' transit operations to private companies, and leaving roads proponents to dominate transport policy debates.

Spatial planning in Melbourne since the 1980s has been characterized by strong rhetoric about the need for a more compact urban form and for greater reliance on alternatives to the car, but budget allocations have consistently favoured freeway construction (Department of Infrastructure, Victoria, 2002). The only attempt to include any significant community engagement in the preparation of spatial plans took place in the 1990s but ended in farce when the clear desire of the public for greater emphasis on alternatives to the car was ignored by planners (Mees, 2003).

Now several decades later, Melbourne is still in a weak position to respond to growing demands for improved public transport. It remains to be seen whether civic action groups will be able to use the growing urgency of the global and local impacts of climate change or other public policy imperatives to create a new window of opportunity for reform of transport policy and practice.

Unlike Melbourne, transport policy in Perth was the subject of only minimal political contention until the late 1970s. Planners, politicians and most of the public shared the view that the car was the future and the railways would die. Announcement of the first rail line closure in 1972, coming as the first oil shocks were being felt internationally, triggered extensive civic protests. Within four months, civic groups, supported by but not beholden to the unions, had organized a huge community campaign. Its strength came from its strong cadre of 'grassroots' organizers, most of whom were new to community advocacy. Nevertheless, even with opinion polls running at 82 per cent in favour of retaining the train, the line was closed.

When a reformist government was finally elected in 1983, rail closures were not a prominent issue. Even so, the new Premier told his ministers privately that he expected significant changes to public transport policy. The ambitious new transport minister, whose strength lay in his abilities as a 'deal maker', was most interested in finding cost-savings for mining and farming interests, and that meant job cuts in rail freight. The rail unions resisted, but the Minister brokered a deal involving an inventive redundancy package for workers and government support for urban transit improvements.

Community rail advocates were allowed a place in policymaking, and their skilful work inside and outside the formal political processes was vital in keeping the government to its commitment to electrify the existing system and, some years later, to build a new rail line to the northern suburbs rather than the busway favoured by many bureaucrats. However, decisions to fund new rail projects would not have been made without an increased acceptance by a senior transport policy professional, Stuart Hicks, of the importance of transit to the future of the city. As a senior transport planner during the 1970s, Hicks had been an active member of the 'roads consensus', but he later came to see the essential place of transit in the future of the city. In 1984, Hicks took on the job of running transit operations in Perth. From this point on, he saw that his main task was to get 'enough people to suspend disbelief and back you to make something happen' (Stone, 2009b: 430).

He introduced common branding of all buses and trains and one set of timetables for all services. Even before rail electrification in 1991, these new services slowed the decline in per capita transit use. And, Hicks argues, if his organization had not become 'a policy influencer ... a marketer and a [successful] manager of big infrastructure ventures' (Stone, 2009b: 430) there would have been no driver to bring the desired system to reality. He was also central in the inclusive planning processes that eventually saw construction of railways beyond the suburban fringe to both the north and south of the city, which more than doubled the length of the suburban rail network.

Despite these significant changes in Perth's transit system, the city is still very car dependent. Freeway construction continued throughout the period in which the new directions in transit planning were established. However, growing popular and professional recognition of the need to limit growth in car use has seen a gradual slowing of the roads programme. Road-planning decisions are not shaping Perth as they did in the past.

Perth has also faced difficulties in creating and implementing a strategic vision for land use and transport in the greater metropolitan area to replace the 'roads consensus' of the 1970s. The city's first strategic plan, in 1955, had envisaged a relatively compact form with a strong role for public transport, but, over time, proposed arterial roads became full-scale freeways and the intended rail expansion never happened. By the late 1980s, some planners were willing to acknowledge the negative consequences of forecasts of a doubling of traffic in Greater Perth (Stokes and Hill, 1992: 124). Their 1990 strategic plan (DPUD, 1990) called for an expanded rail network into the growth corridors, and containment of growth within an urban boundary. The plan contained detailed policies for traffic demand management, and provision for walking and cycling. However, public engagement in the development of the plan was minimal. With no public or political 'buy-in' for this shift in priorities, the freeway rollout continued and further expansion on the suburban fringe inevitably followed.

Learning from this experience, an 'activist' Minister in a new centre-left government, elected in 2001, brought transport and land use planning into a single agency and began a programme of deliberative community engagement (Government Media Office, 2001). This process was, by Australian standards, quite rigorous in its attempts to engage with 'hard to reach' groups including the indigenous community, young people and those whose first language is not English. A large one-day community engagement workshop and a 12-month process of translating community sentiment into policy objectives were used to underpin a new strategic spatial plan. This plan (WAPC, 2004) supported investments in new rail projects, but the consensus on its broad directions among planners and politicians was still not strong. In fact, a new metropolitan plan (WAPC, 2010), prepared using a more traditional plan-making process with less community engagement and weaker restrictions on suburban expansion is now in force.

Unlike the Australian cities, power over urban affairs in Vancouver is shared between the province and local authorities. Importantly, since 1960 the Greater Vancouver Regional District (GVRD) – now called Metro Vancouver – has provided a structure for cooperation among councils and a strong resource base for municipal politicians from which to organize during regular periods of contention with the province.

Vancouver's early streetcars and 'interurban' light rail shaped the city's form, but by the 1950s, this system was in decay and – to government officials – freeways were the obvious replacement. There was no established urban road construction agency, but Vancouver Council staff hired American consultants who prepared a large-scale plan typical of the times, and moved, behind the scenes, to assemble resources for construction from federal, provincial and local sources.

When the plans became public in 1967, huge opposition spawned a new political party, which won control of Vancouver City Council in a landslide in 1972 and acted quickly to wrest power away from the bureaucracy by sacking the City Manager and five other department heads, including the planning director, and five of the eight other planning staff. New and surviving planners were keen to work with the new agenda 'without creating a crisis ... among developers and architects' (Leo, 1994: 678). Thirty years on, their effort is celebrated as the 'Vancouver Achievement' (Punter, 2003).

The newly formed GVRD also became a location for innovation (Lash, 1976). Its decision to take a consultative approach to regional planning, although 'hard fought' (ibid.: 55), produced long-lasting policies for growth control and anticipated a much greater role for transit in the urban transport system (GVRD, 1975). Offered more funding, the transit operator achieved solid patronage growth, partly because its system operated as a coherent network.

This success helped to build political support for transit expansion that took more than a decade to come to fruition, and then only out of intense political conflict (Berelowitz, 2005; Punter, 2003). In the competition for the Expo 86 world fair, Vancouver needed transit investment. Conservatives favoured an elevated, driverless light-rail system. Their opponents preferred an existing plan for conventional light rail. Eventually, the elevated, driverless Sky Train became the centre of a working transit system, not just an event toy.

Following a conservative electoral backlash during a recession in 1983, supporters of growth control and transit recognized the need to renew their political mandate. The GVRD set up a public process including conferences, public opinion surveys and a wide range of work in political and technical committees to evaluate various planning options. These eventually produced a new land use blueprint, the *Liveable Region Strategic Plan* (GVRD, 1996) and a remarkably coherent transport plan, *Transport 2021* (GVRD, 1993), that set out a detailed programme to improve transit and to allow road congestion to grow as a demand management tool, particularly on the bridges over Burrard Inlet on to the downtown peninsula.

The GVRD is now justifiably proud of both the content and the process that delivered these plans. However, their beginnings, again, were contested. Politicians were sceptical of the value of processes that were open to the public at an early stage and were not controlled by 'expert opinion'. As in 1976, the GVRD resolution to begin a public process was passed by just a single vote.

The GVRD planning process included strong engagement with the public, but unlike Perth, where the community was the primary focus of the consultative effort, politicians and planners were also directly involved. This was because of the history of contention over planning policy between the province and the region dating back to the 1970s. During a period of intense conflict over suburban development in 1983, the Province had removed the GVRD's statutory planning powers. So, to take practical effect, a new metropolitan strategy would require the support of all 22 municipalities (Harcourt and Cameron, 2007). The process of deliberation and negotiation to achieve this took seven years, but it produced a plan that was owned by local government planners and politicians. Indeed, it survived for 15 years as the region's strategic spatial strategy. Only in 2011 did Metro Vancouver replace the 1996 plan with a more contemporary version. This new plan is consistent with previous plans, and was adopted unanimously by local governments.

Conflict continues in Vancouver between supporters of road construction and those who wish to strengthen urban consolidation policies through further transit investment and the so-called 'eco-density' agenda (City of Vancouver, 2006a). Even so, new transit infrastructure continues to be rolled out, growth controls remain in force, and transit ridership continues to grow.

Theorizing success and failure in changing transport policy

In the case study cities, it is clear that change, when it did occur, did not follow from any single moment of new 'policy learning'. Instead, there is a continual process of contention, which we can call 'challenge and resistance', between the supporters of different conceptions of transport policy. Obviously, decisions on programmes and expenditure are made by those with the power to do so. Therefore, this struggle for power must be analysed and understood.

In Chapter 5, it was argued that a global shift in the regulation of capital, which took place with the collapse of Fordism in the 1970s, provided opportunities for the growth of mass movements for improved urban amenity that we have seen in all the urban regions described in this chapter. While this is true, it is clearly apparent that not every urban region achieved a homogeneous transport network under Fordism, and not every city lost it under neo-liberalism. The political trajectories of the cities described above do illustrate how important transport systems are as triggers for rupture or tools to contain dissent against the power of business and government. The differences between the individual cities – the success in some and failure in others – are crucially important, and they allow us to theorize about the local conditions required for transport policy change to occur.

It appears that a fundamental shift in transport policy in an urban region will occur seldom, if ever, without significant change in the membership of the group with power over policy directions. Achieving such change is clearly very difficult. Change is rare, but it does occur, and both the empirical evidence from the case studies and a wide range of theoretical perspectives point to the importance of the political dimension in understanding how power is gained and maintained. What can the theory of action tell us about the difficulty of change and the conditions for it to occur?

For many, from both the right and the left, the shape of our cities is overwhelmingly the result of economic forces. However, when trying to explain variation in cities across the globe and within specific regions, economic analysis, such as that in Chapter 5, can be rich in description and categorization, but ultimately it lacks explanatory power. Stronger explanations of observed variation in urban outcomes can be achieved when the analysis recognizes the role of local politics as a mediator of apparently irresistible global economic forces. An example of this is Leo's investigation of the causes of variation in urban planning outcomes in Vancouver and Edmonton. Leo showed that Vancouver's superior performance was less a result of its comparatively higher placing in the global economic hierarchy than the product of a 'substantial exercise in state power' (Leo, 1994: 690) that imposed 'significant constraints on developers and [gives] a serious measure of influence ... to citizens, citizens' groups, and the local state' (ibid.: 683).

A valuable explanatory bridge that posits a mediation of the impact of the global economy through local governing arrangements is provided by 'regime theory'. This theory was pioneered by Clarence Stone (1998, 1989) as a means to understand his observations of the surprisingly peaceful trajectory of economic and social policy change in Atlanta through the years of the civil rights movement and beyond. Stone argues that structural economic power is not enough to explain how 'in the face of complex and ... divisive forces, an effective and durable capacity to govern can be created' (1989, p. xi). His key insight was that, while business elites have access to significant resources, they do not exercise command power. Instead, in order to bring significant projects to fruition, the capacity to do something is created by informal arrangements between public bodies and private interests. Stone defined a 'regime' as the set of actors who build the relationships to maintain such arrangements over a long period. He calls such a group of actors a 'governing coalition'.

For our purposes in understanding the strength of resistance to change in urban transport policy, Stone's characterization of urban governance as the 'power to' rather than 'power over' is useful. He demonstrates that, even when 'citizens are aware that the community could be organized on some basis other than investor prerogative' (Stone, 1989: 222), a regime can engender cooperation through agreement on purpose – even if this is constrained by a 'bounded rationality' (Stone, 1998: 256). And, it can maintain its power through 'selective incentives' and 'small opportunities', which once established exert a 'gravitational pull' that puts significant obstacles in the path of any challenge group. This theoretical position is consistent with our observations of local contention over transport policy in which various conflicting policy trajectories are clearly articulated, but only one grouping can assemble the means to put its policies into practice: usually through control of significant public investments.

These governing coalitions can be built around progressive agendas such as sustainable transport policy, as we have seen in our Vancouver case study. Another progressive regime has existed over several decades in Portland, Oregon. This has had some success in promoting efficient land use and environmental safeguards across an entire urban region through a 'long series of remarkable political developments' (Leo, 1998: 368), although its sustainable transport achievements are not as remarkable as those in Vancouver. It is made up of businesses that accept that their interests are served by managing growth intelligently, farmers who want to maintain viable commercial agriculture in a region with limited arable land; and environmentalists, who, while not considering themselves as 'insiders', have been 'powerful and central actors' (Leo, 1998: 381).

As we have seen in Vancouver and Perth, Leo observed a pattern of informal networks of cooperation and influence. The environment groups that have played a major part in the coalition – notably the *1000 Friends of Oregon* – have been stable and consistent, something very rare among community organizations. Leo characterizes their leadership as 'exceptionally perspicacious in understanding political manoeuvring', able to exploit opportunities for leverage between different levels of government, and 'self-conscious about shifting back and forth between a confrontational strategy designed to expand the debate and a coalition-building approach designed to build a majority' (Leo 1998: 381).

So, we can see that there are opportunities for local politics to mediate the impacts of local and global economic structures and shape the transport policies of an urban region. A number of different perspectives from political science and from the analysis of public policy processes can help us understand the ways in which the dynamics of these local political processes have shaped urban transport policies in more sustainable directions.

The fragmentation and complexity of modern governance observed by the regime theorists has also changed the way that scholars look at finer levels of the policy development process. In a move away from studies of formal institutions (like that of Anderson, 1975), attention has shifted to the patterns in interactions between members of what came to be known as 'policy networks' – groupings of specialist actors such as bureaucrats, legislative personnel, interest group leaders, researchers and reporters (Campbell et al., 1989; Richardson and Jordan, 1979; Sabatier, 1991). From this perspective, it is the mode of interactions that is important: the networks are not the actors or their institutions but rather a 'kind of capillary system through which work is done (and the) means for extending or modifying the paths available for the next expected encounter between regular participants' (Considine, 2005: 140). But, equally the network is often the site for strong resistance to change, and this resistance can be carried out through a 'black polity' with 'routines ... that do not derive their legitimacy from the public processes and values of parliamentary democracy and might ... circumvent these' (Raab, 1992: 74).

Some attempts have been made to use the concept of policy networks as a means to develop explanatory theories about ways that their role in policy inertia might be circumvented (Heinz et al., 1993). However, such approaches can be complex in their data requirements or superficial in their inability to distinguish between government players operating as disinterested intermediaries or as the captives of powerful interest groups. In any case, it is the character and action of the individuals within a network or a challenge group, rather than the networks themselves, that determine whether change takes place. So, as Dowding (1995) suggests, it is more useful to think of the policy network as a powerful metaphor and descriptive device for the forces that create resistance to policy change. In this light, it is useful to observe, within the range of possible policy networks, that a distinction can be drawn between a 'policy community' – where small numbers of interdependent professional actors maintain strong and persistent control over a particular policy sector – and larger, more open 'issue networks' where conflict is present and outcomes are more in doubt.

Adding to the complexity of the study of the behaviour of policy networks, individual members of these networks typically work within institutions that operate according to rationalities that are not always obvious to outside observers. Institutional responses to internal or external pressures are complex and unpredictable and do not conform to simple exchange theories. In fact, 'the ability to create change ... does not guarantee either that any arbitrary change can be made at any time nor that changes will turn out to be consistent with prior intentions' (March and Olsen, 1996: 256). Therefore, in order to determine causal factors for the success or failure of any serious attempt to change urban transport policy it is necessary to have a thorough understanding of the local institutions for transport management and for urban governance.

However, even though success or failure of attempts to change urban transport policy might be determined by local institutional factors, some more generalized theories of the policy process point to the existence of 'windows of opportunity' in which change is possible. Analysts have observed three 'streams' of actor behaviour and process within policy networks. These are:

> a *problem stream* consisting of data about various problems and the proponents of various problem definitions; a *policy stream* involving the proponents of solutions to policy problems; and a *politics stream* consisting of elections and elected officials.
>
> Sabatier, 1999: 9

Under this model, opportunities to escape the control of an existing policy community and to begin the task of creating new networks organized around a different agenda come about when political entrepreneurs are able to 'couple the various streams' (ibid.: 9). It is at these times that opportunities exist to overcome problematic 'path dependencies', discussed in Chapter 6.

What the case studies tell us

Consistent with these theoretical understandings, the case studies show that local politics was central to delivering improvements in sustainable transport. In the 'successful' cities, political entrepreneurs seized opportunities to construct new policy communities to shape and implement programmes, and community advocacy played an essential role in creating these opportunities. The skill and effort required by political entrepreneurs and community advocates is considerable, and the margin between success and failure is slender.

In Melbourne, as predicted by regime theorists, challengers could not build sufficient support for the task of re-orienting urban planning and transport policy towards containment and transit. This was a much bigger 'purpose' than the construction of a series of freeway projects for which the justifications were carefully limited to simple claims of the economic benefits of improved local traffic flows. Limited access to resources was also a problem for challengers in Melbourne. In Vancouver, on the other hand, the pro-road suburban developers found themselves without access to institutional resources.

Another failure to turn rhetoric into action – not included in the case studies, but still worth noting – is the trajectory of British government policy on inter-urban motorway construction in the decade from the early 1990s. It provides clear illustration of the importance of Bratzel's second condition for creating new policy communities: new people must be installed in important political and bureaucratic positions.

From around 1990, strong social movements emerged to challenge the Tory government's massive road-construction programme – promoted as the 'biggest since the Romans'. Alliances between environmentalists and middle-class landholders, and a change of government, helped give legitimacy to professionals who, over many years, had challenged the dogmas of the road-builders. A programme for change was proposed, under the slogan of 'new realism' (Dudley and Richardson, 2000; Vigar, 2002). However, new political leadership, new policy rhetoric and a strengthened 'issue network' were not sufficient to displace the entrenched policy community. As Sloman (2003) amply demonstrates, the road-builders, who had remained in positions of power and influence, were able to exploit local political events to counter the influence of the sustainable transport lobby and turn the new political leaders back to the old policies of automobility.

As would be expected, business interests hold influential positions in the governing coalition of all urban regions. However, the evidence from the case studies suggests that, while shifts in the allegiances of some business interests were necessary for new governing coalitions to come to power, new directions in transport policies were the work of politicians, professionals and community advocates. Their leadership was the key to winning support from business leaders, not the reverse. Hicks' comment that, in Perth, he was asking people 'to suspend disbelief' while he attempted to rebuild the transit system is a good illustration of this point.

There is great variation in the way the jurisdictions of various levels of urban governance overlap and in the way that functional responsibilities for the management of transport systems are distributed. Varying jurisdictional structures shape the range of strategic and tactical options open to those who are attempting to bring about policy change. But, the case studies show that significant change is possible in a variety of structural and institutional patterns.

The centralization of control over urban policy and resource distribution held by the Victorian State government, and its structural dominance over weak local governments in Melbourne, contributed to the lack of success of transport policy challengers; while in Vancouver, the more equal distribution of functions and responsibilities gave challengers more chance of success. Supporters of sustainable transport were able to build an institutional base in the GVRD and at the City of Vancouver that survived when their opponents held sway at the provincial level. However, the experience in Perth, where structures of urban governance are broadly the same as those in Melbourne, suggests that actor behaviour can overcome limitations in the structure of urban governance.

Consistent with Bratzel's observations, there was a very strong civic challenge in Perth to the plans to close the Fremantle line. Without such a public 'crisis', it is unlikely that political leaders would have had sufficient motivation to create opportunities for new transit policies. However, there was no rapid or wholesale change in the bureaucracy in Perth as there was in Vancouver or in Bratzel's cities. Changes in personnel did occur, but over a longer period. This does not necessarily undermine the validity of Bratzel's general conclusions. Perth's transit system in the 1980s was so small that reform could take place away from the political spotlight. The process of renewing the transit system did not directly threaten road construction programmes and so there was no organized opposition to transit reform from the road-based transport policy network. However, the unusual climate for reform and policy analysis established within the reformist government did force the various protagonists in the transit reform debate to engage in ways that researchers like Sabatier (1991) have observed is most compatible with policy-oriented learning and change. Key issues of transit reform – such as electrification versus diesel railcars, or bus rapid transit versus rail in the northern suburbs – were technically tractable. The protagonists were forced by political circumstance to resolve these issues in professionalized forums for which they were temperamentally and intellectually well suited. Simply, an unusual constellation of events and personalities shaped Perth's transit reforms.

Conclusions

This chapter has demonstrated that deliberate change is possible and that it requires skilled work by community activists, political entrepreneurs and reformist professional staff acting, if not in concert, at least each showing respect and understanding for each other's roles. Creating lasting change requires the ability, first, to exploit the occasional conjunction of problem identification, policy response and political opportunity that allows new directions to be taken, and, second, to implement practical projects that demonstrate the effectiveness of the new directions and recruit new public and political support.

These 'windows of opportunity' will reflect the nature of local political institutions and will not necessarily be predictable, but taking advantage of their appearance requires considerable effort. This effort will take many forms, but its purpose is to build legitimacy for new policy and programme understandings by demonstrating their rationality – or the irrationality of the status quo. This can be done through deliberative processes – both with and without the imprimatur of government; by encouraging and supporting emerging political entrepreneurs; and by forming working relationships with those seeking social change in other fields. It is hard work, but the evidence shows there is no easy way.

9 New analysis for a new synthesis

Matthew Burke, Carey Curtis and John Stone

Introduction

As we argued in Chapter 1, dealing with transport problems leads us to traverse the multiple dimensions of theory, policy, process and applications, including the area of technical assessment. The problems of assessment are significant, with many of the practice technologies in use contributing to the stubborn inertia of automobility (Banister, 2002: 22; Vigar, 2002; 2006). We discussed in Chapter 4 some of the issues surrounding technical analysis, focusing on how the problem of 'congestion' has come to be construed, and the technological obsession with measurement and valuation of time allegedly saved by construction of road infrastructure. As we argued in that chapter, however, there is a strong institutional interest in broadening the scope of transport assessment towards questions of equity, access, oil vulnerability, climate change and other emergent concerns (see for instance Dodson and Sipe, 2007; Hickman, Ashiru and Banister, 2010; Preston and Rajé, 2007).

This chapter explains a new set of practice technologies designed to assist planners and decision makers when confronted with complex transport problems flowing from the dilemma of mobility. The first of these (SNAMUTS) introduces concepts of 'place' and 'node' using a GIS (geographic information systems) technique to model the incidence of both places and nodes. The model demonstrates both network accessibility and opportunities for place-based development, and thus models the potential of different urban structures to deliver transport accessibility. The second models transport patterns via census data – i.e. the jobs-housing balance – assessing public transport and walking accessibility, and quantifying carbon and oil vulnerability impacts (MULUTT). The third applies concepts of public transport network analysis to show how an effective public transport system can be delivered to low-density suburban environments.

A user-friendly approach to the development of planning tools

Our general approach to the development of new practice technologies has been to recognize that:

- transport policy decisions are inherently political
- most transport planners remain wedded to a technocratic role in advising decision makers
- models – inputs and outputs – have not generally been developed in a discursive setting where stakeholders can participate fully in the decision processes (further discussed in Chapter 11)
- transport planning has mostly moved away from comprehensive approaches towards a focus on specific projects and policies (Banister 2002: 131), and
- the addition of new tools that unpack new dimensions and realize alternative possible 'solutions' is a fruitful way of shifting the balance towards sustainable transport.

This approach overcomes any defensive reactions from practitioners regarding their existing applications and processes, providing a way forward they can relate to and engage with productively.

In the modelling work, our intent has also been to provide advice and graphic/spatial displays comparable with those employed by existing modelling frameworks. This allows new practice technologies to be more readily adopted and added to existing frameworks of analysis. This approach also allows outputs to be provided that are, for the most part, consistent in format and structure with the outputs of earlier more conventional modelling.

Much of our work has been in the realm of applications developed within GIS. Many of the ideas developed are not new, but the tools presented below have harnessed the power of contemporary spatial analysis and made significant advances in the ways in which transport problems may be assessed in cities.

Spatial network analysis for multimodal urban transport systems

It is clear that, in policy terms, planning for accessibility is a key requirement. One of the enduring issues, however, has been how to turn these aspirations into mainstream planning practice. There are examples where accessibility tools have been developed and applied but to date these applications have tended to be single-issue focused. There remains a gap in accessibility tool development for a multi-focus perspective measuring accessibility from every centre to every other centre in a city-wide application. In addition there is also a strong need for accessibility tools that can enhance the public understanding of land use-transport integration. To achieve these objectives, the inputs and outputs of accessibility tools need to be communicated in a way that can be easily understood by professionals, stakeholders and the informed public. The Spatial Network Analysis for Multimodal Urban Transport Systems (SNAMUTS)[1] tool has been designed with these needs in mind.

The tool is intended to aid discussion and to lend weight to decision making within the fields of land use planning and transport planning, particularly where outcomes leading to more sustainable transport options are needed. The development of the tool has been closely linked to needs arising from the recent Australian metropolitan planning strategies: *Melbourne 2030* (Department of Infrastructure Victoria, 2002) and Perth's *Network City* (WAPC, 2004) where there are movement and activity centre networks promoted as redevelopment corridors and nodes. While such strategies articulate key sustainable transport concepts such as the desire to promote public over private transport, and land use-transport integration, to date the basis for critical decisions on which centres should be promoted, or where to improve the public transport network, have been made in a vacuum.

SNAMUTS is a GIS-based tool designed to measure the accessibility provided by existing or proposed urban public transport networks in their current or proposed land use context. In particular, SNAMUTS endeavours to identify and visualize in a coherent mapping exercise a land use-public transport system's strengths and weaknesses of geographical coverage; the ability and efficiency to connect places of activity; the strategic significance of routes and network nodes; and the speed competitiveness between public transport and car travel. Careful consideration is given to how the analysis of public transport networks diverges, and requires different assumptions and definitions, from the analysis of movement networks for individual transport – i.e. pedestrians, cyclists and motorists.

SNAMUTS was developed following a review of a range of common accessibility measures. There is a considerable diversity in ways of measuring accessibility including: spatial separation measures, contour measures, gravity measures, competition measures, time-space measures, utility measures and network measures (see Curtis and Scheurer, 2010 for a fuller discussion). This diversity indicates that there is no single perfect accessibility measure; rather, it is critical to apply several measures in combination in order to present the necessary information to inform debate and deliberation. Each measure provides useful insights with which to assess the performance of a public transport network and its correlation to the land use structure and geographical distribution of activities.

The origins and evolution of SNAMUTS were inspired by the node-place dichotomy posited by Bertolini, the Space Syntax model, and some more recent contributions on the context between urbanism and transport networks (Porta, Crucitti and Latora, 2006a, 2006b). Bertolini's 'node-place model' (Bertolini, 1999) is critical to the understanding of accessibility within the context of planning for land use-transport integration (LUTI). He notes that while railway stations provide access to the transport network (the 'node'), they also offer a 'place' function. That is to say the station precinct can also be a destination where land use activities are available to public transport users and others. This is an important dimension – the accessibility of *opportunity*. This an important aspect of quality of place mentioned in Chapter 4. Therefore, in these terms, accessibility means more than simply the accessibility of the network in the sense of the ease of getting to a place. For instance, a network node such as a transit station may be highly accessible on the public transport network, yet the surrounding site may be occupied by car parks or obsolescent industrial buildings so the land use opportunity to make it also an attractive place is lost.

Within SNAMUTS, accessibility by public transport is measured from the perspective of a traveller who also has the alternative choice of car travel – 'choice user' – to any location within the metropolitan area. This is an important way of conceiving of accessibility since, in Australian and North American cities, most households have access to a car and even poorer households depend on cars for mobility. Therefore if public transport is to compete with the car in any assessment for future planning, it must be measured from this perspective. European transport planning already commonly takes 'competing with the car' as best practice for public transport (see Nielsen et al., 2005: 22).

Ease of movement, a common definition of accessibility (Bhat et al., 2000), is measured using an impediment variable that deploys three factors critical to the decision to use public transport: inter-peak travel time by public transport, frequency of service, and the transfer time between public transport services. Again considering the 'choice user', the possibility of undertaking multi-purpose journeys from any centre to any other within the metropolitan area is factored in.

With the above conceptual framework in mind, seven SNAMUTS indicators are then applied, highlighting network performance from several angles:

1 *Degree centrality* by transfers – or topological centrality – measures the average minimum number of transfers required for all network paths, and from each node to reach any other node on the network.
2 *Closeness centrality* by impediment – or metric centrality – measures the average minimum cumulative impediment – travel time divided by service frequency per segment – for all network paths, and from each node to reach any other node on the network.
3 *Efficiency change* following network modifications – actual or proposed – measures the before and after ratio of inverse cumulative impediment for all network paths, and from each node to reach any other node on the network, and is weighted by combined activity node size.
4 *Contour catchment* measures the number of residents and jobs in activity nodes accessible within 30 minutes from the reference node, and the average size of all 30-minute contour catchments across the network.

5 *Betweenness centrality* measures the percentage of all preferred network paths[2] – weighted by cumulative impediment and combined activity node size – that flow through each route segment.
6 *Speed competitiveness* measures the ratio of lowest available public transport travel time with road travel time in congested conditions for all network paths, and from each node to reach any other node on the network.
7 *Network connectivity* measures the propensity of each activity node to act as a transfer hub – and thus attract potential activities to service transferring passengers – and includes a cumulative count across the network.

Finally, a composite indicator for overall public transport accessibility was compiled from the figures for each activity node in degree centrality, closeness centrality, contour catchment, 'betweenness' centrality (measured at the node), speed competitiveness and network connectivity (Curtis and Scheurer, 2010).

There have been several applications of SNAMUTS. The first was in demonstrating the city-wide changes in public transport accessibility following the opening of the Perth–Mandurah rail line (see Chapter 8 for a discussion of the political processes leading to the new railway). Not only was there a new railway but the public transport timetable was also substantially modified by re-organising bus services so that they were integrated fully with the new rail service in order to meet the objective of providing for 'seamless travel'.

One of the significant outputs can be seen at Figure 9.1 (seen in colour in the Companion Website for this volume; Figure 9.1), which uses the composite indicator to map public transport accessibility using a 'traffic light' colour code: green shows the population/employment catchments with best potential accessibility, while red shows those that are the least accessible. Black is used to denote those areas where the public transport service was below even the minimum standard – two buses/trains per hour. These maps showed that the outer suburbs were the least accessible, although there were variations among suburbs. In particular, some areas within the central city measured quite poorly from an accessibility perspective. This finding was reported in local newspapers and demonstrated to us the value for the general public of mapping outcomes in this particular way. Thus people could now see how accessible their locality was, and, on the basis of this evidence, could go on to argue for improvements to public transport.

The SNAMUTS tool has resulted in an enhancement of existing practitioner knowledge and a better understanding of land use-transport integration. By measuring accessibility before and after the new railway (Perth-Mandurah), the value of investing in new public transport infrastructure was demonstrated, shown in the mapping where catchments turned from red to green. Indicator 3 showed that there was an improvement in

Figure 9.1 SNAMUTS Composite indicator showing changes in potential accessibility by public transport after a new railway was opened in Perth

'efficiency' of 32 per cent. Not only was public transport accessibility measured for the first time, but it was also measured in a language that appealed to professionals, themselves more used to working with traditional transport models, at the same time as being presented in a way that was readily understandable by the public.

The mapping prompted a much clearer understanding of LUTI – it showed that one station on the new railway line (Bull Creek) had gone from orange to green and was now second only to Perth CBD in terms of accessibility. While this was of interest from a transport network perspective, more importantly it highlighted the mismatch between node accessibility and actual land use around the station. Bull Creek station is currently a transit interchange surrounded by major road infrastructure and car parking. After an investment of AUD$1.9 billion in the development of the new railway, this new understanding raised questions about why a higher intensity of land use had not been planned to integrate with the station and capitalize on the improved accessibility. This example demonstrated what LUTI might actually mean for implementation. Up to this point LUTI had in-principle support as a policy ideal, but there had been no evidence demonstrating where or how to translate this policy ideal into action.

Modular urban land use and transport tool

Taking a slightly different approach, another project has also been attempting to improve conventional methods of transport-land use analysis and evaluation to support the development and implementation of sustainable transport systems. This project has focused on supplementing conventional 'four-step' transport modelling (see Box 9.1) with additional approaches to deliver greater awareness of and insight into the potential impacts of specific transport and land use scenarios.

The project is developing a new modelling framework, the 'Modular Urban Land Use and Transport Tool' (MULUTT). Modules for exploring transport patterns via census data – i.e. the jobs-housing balance – assessing public transport and walking accessibility, and quantifying carbon and oil vulnerability impacts, are in development by the research team. The intent is to use similar, if not identical, inputs to those used in conventional four-step models, including descriptions of the road and public transport network, land uses, population, and employment, to reduce the volume and complexity of necessary data collection. As with SNAMUTS, the MULUTT modules are mainly GIS-based. The full framework, and its sub-components, are now being harnessed to assess transport and land use scenarios. They are proving particularly useful for exploring broader 'What If?' questions for long-term urban restructuring.

MULUTT modules have been adapted for use alongside the *Brisbane Strategic Transport Model* (Multi-Modal) (BSTM), the main transport model used in Brisbane, which was obtained under license from the Queensland Department of Transport and Main Roads (TMR) for the project. However, the accessibility module has been used in other areas,

Box 9.1 The four-step model

The traditional strategic transport modelling approach for large cities is often called the four-step model as it involves the analytic steps of:

1. Trip generation – how many trips there will be.
2. Trip distribution – where those trips are going.
3. Mode assignment – which mode of transport will be used.
4. Route assignment – which particular routes will be taken.

The four-step model normally uses data on past travel behaviour, projecting the pattern of the past into the future. It could be called the classic 'predict and provide' model.

Four-step models are used for scenario testing, calculating predicted travel times and flows across transport networks. These models assist planners in comparing possible projects, and different transport and land use arrangements.

most notably in exploring public transport accessibility to proposed stadium locations on the Gold Coast – Queensland's second largest city (Burke, Evans and Hatfield, 2008; Burke and Woolcock, 2009). This application of the project was a useful trial of the power of an 'alternative' view of a transport and land use problem.

The Australian Football League (AFL) recently built a new stadium on the Gold Coast to seat more than 40,000 fans. Most contemporary stadium developments are transit-oriented, and implement extensive travel demand management measures to curtail car access and promote the use of public transport by patrons. Accessibility by public transport is a critical planning dimension, though this appeared to be less of a consideration for the sports bodies involved in the planning of this particular stadium. Prior to a location being selected by the AFL, the research team used the MULUTT accessibility model to assess four of the key sites being considered, as well as the competitor Skilled Park rugby league/soccer stadium site (Figure 9.2).

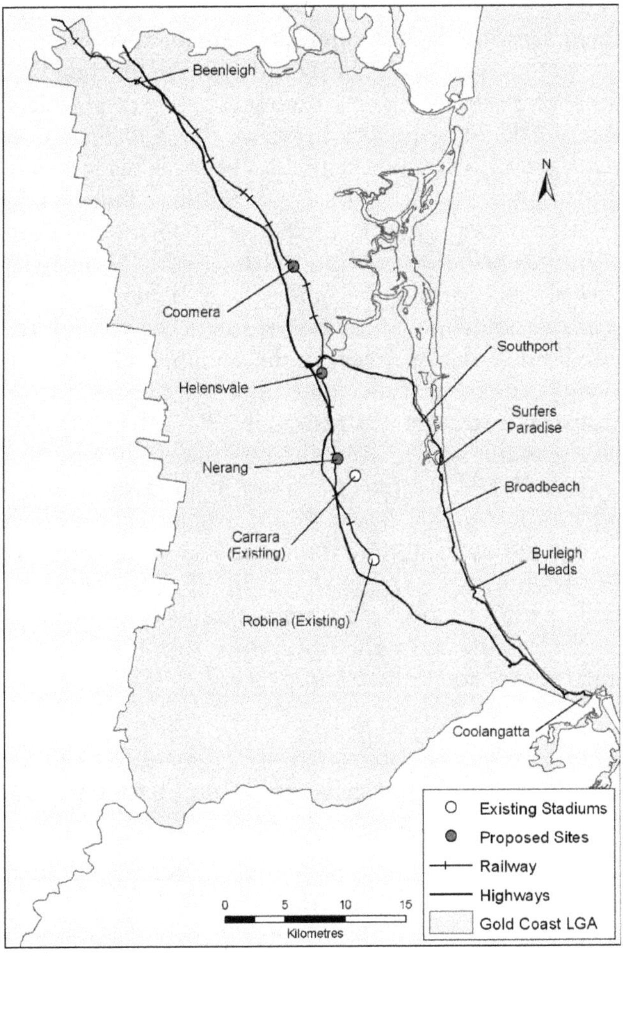

Figure 9.2
Existing and proposed stadium locations

The analysis sought to determine the number of Gold Coast residents who could access each of the proposed stadium sites easily by public transport. Ease of access was defined in terms of travel time, considering all the sectors of the public transport journey. This included time spent either walking to and from transit stops, or interchanging between services. The intention was to help the sports bodies in selecting a transit-oriented location that could better serve the community.

The methodology is explained at length in Burke et al. (2008). Essentially, the approach focused on all stages of a journey from home to the stadium, including:

- walking access from home to a public transport stop;
- waiting time for the public transport vehicle;
- travel time on the vehicle;
- interchange with other public transport services (if necessary); and
- the walk from the final public transport stop to the proposed stadium site.

Catchments for each stadium site were mapped and populations were highlighted that were accessible within 45 minutes total travel time – defined as 'high' accessibility – and 75 minutes total travel time – 'reasonable' accessibility. Spatial mapping of ease of access to the site eventually selected a site, Carrara, shown in Figure 9.3 below. This shows that the stadium site, located away from the city's only rail line, failed to provide high levels of access to the key growth corridor of Coomera and Beenleigh. Comparisons with other locations on the rail line, such as at Helensvale and Nerang, showed that these sites could service a much greater proportion of the existing population, and also the growth corridors. Locating at Carrara may have been to the detriment to the stadium, its tenant club and the sporting code of Australian rules football. Alas, despite evidence-led advocacy, the cheaper Carrara site was eventually selected, partly due to the emergence of the global financial crisis at the very time a decision had to be taken.

The MULUTT project is now focused on exploring issues of urban decentralization. It is examining the transport impacts of government sector employment relocation in Brisbane, work that will be reported shortly (Burke, Li and Dodson, 2010). This work is helping the team to recast existing modelling frameworks to demonstrate the value of revised approaches, to apply the framework to the BSTM, and to advocate that practitioners adopt these approaches.

A benchmarking tool to support planning of new public transport service patterns: applying European 'network' planning to Australasian cities

Recognising both the political and technical dimensions in transport policy decision making, the third project described in this chapter was built around engagement with transport planners in New Zealand's three largest

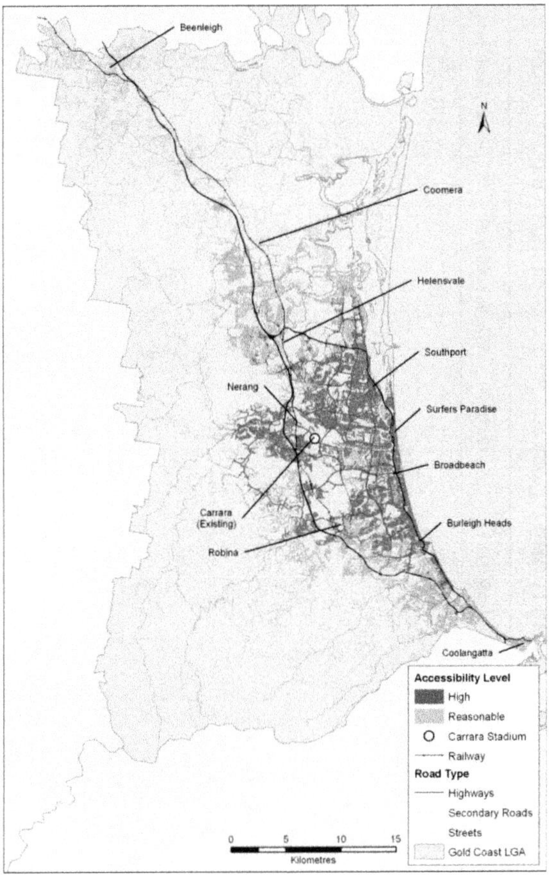

Figure 9.3
Public transport accessibility to Gold Coast Stadium, Carrara

cities. It sought to build institutional and political support for restructuring of public transport service patterns in ways that reflect the experience of successful operations in specific European and North American cities.

For this project, a tool for benchmarking of public transport performance and service supply parameters was developed. Its purpose was to demonstrate the potential for more effective use of existing resources and to focus attention on practical changes that local transport planners and operators could readily implement. Its importance lies in its demonstration that good public transport service need not be limited to cities of high population density.

The results of the benchmarking exercise are set out in Table 9.1. The comparator cities were chosen for their similarity in urban form and available transport infrastructure to one of the New Zealand cities. Importantly, reliable data on travel patterns and public transport operating conditions were available from census authorities, public transport operators and from the Kenworthy databases (Kenworthy and Laube, 2001).

Table 9.1 Urban form and public transport performance in the three New Zealand urban regions and their chosen international benchmarks

	Auckland	*Vancouver*	*Wellington*	*Zurich*	*Christchurch*	*Schaffhausen*
Total population (millions in 2009)	1.3	2.1	0.45	1.3	0.35	0.044
Density (persons per ha. in the urban region)	18.9	17.1	22.0	37.6	17.0	36.7
Urbanised area (km^2)[a]	690	1230	205	345	206	12
Jobs in CBD (% of total jobs in the urban region)	13.5	12.6	22.0	12.2	16.6	n/a
Passenger cars/1,000 residents	523	519	489	513	567	530
Car travel cost index[b]	4.2	2.3	4.2	2.6	4.2	2.3
Share of work trips:						
Public transport	7.0	16.5	17.1	40.7	5.2	40.7
Walk and cycle	5.6	8.0	13.4	12.1	12.3	14.9
Car	87.4	74.4	69.6	47.2	82.5	44.4
Public transport boardings (millions of 'unlinked' trips)	52	283	34	542	16	13
Boardings per capita	40	135	77	417	46	289
Public transport service-km (bus- and tram-km plus train-km: millions)	42.3	116.2	24.5	71.9	18.8	2.7
Service-km/capita	32	55	61	57	59	61
Subsidy per boarding[c]	2.54	1.10	1.70	0.45	1.62	1.05

Notes

a For all urban regions except Zurich, these areas are defined by the boundaries of contiguous urban development. This is typically much less than the area within the relevant administrative boundary. In Zurich, the urbanised area is dispersed within rural land and forest over a total area of 1840 km^2.

b This index measures 2009 average petrol prices relative to the median individual wage.

c Equivalent in $NZ at October 2009 exchange rates.

As expected, public transport's share of work trips and per-capita tripmaking rates were much higher in the international comparators. However, some aspects of the data would be surprising to many conventionally trained transport planners and economists:

- The difference in per-capita tripmaking was greater than the difference in mode share for work trips: for example, mode share for work trips was 2.5 times higher in Zürich than in Wellington, but tripmaking was more than five times higher.
- Better performance in the comparator cities was achieved for similar investment in service per kilometre as is currently provided in the New Zealand cities.

These observations were used to direct the attention of local planners to the success of the 'network' approach to public transport planning in North American and European cities. Network planning is a means to provide the public transport user with the convenience of car travel at an affordable cost to the public transport operators. It relies on the provision of a carefully planned system – the *network* – that uses reliable and convenient transfers between services to allow users to choose their own travel path.

The network concept was identified by Mees (2000; 2010), who has demonstrated that its application is a more accurate predictor of public transport success than the density of residential settlement (as argued for instance by Newman and Kenworthy, 1999). This is supported by a comparison of public transport outcomes in a significant sample of US cities (Brown and Thompson 2008), which showed that the network approach produced comparatively higher patronage growth and lower rises in subsidy levels than conventional service planning paradigms.

The hypothetical example of 'Squareville' (Figure 9.4) illustrates the key idea behind network planning: serving the most journeys with the least resources. It shows that transfer-based systems can greatly exceed the low elasticities of demand used in traditional transport studies. The network effect enables increases in patronage at a faster rate than resources are added.

Unfortunately, the only material available, in English, for public transport planners on the practical application of the network approach are contained in the Vuchic's 2005 textbook *Urban Transit: Operations, Planning and Economics* (Vuchic, 2005) and Nielsen's *HiTrans handbook for practitioners in small to medium European cities and towns* (Nielsen et al., 2005).

This GAMUT project addressed this gap through a series of seminars designed to provide local planners, politicians and communities with simple descriptions of the key elements of the network approach, namely:

- a simple line structure for ease of use and reduced resource requirements;
- stable line and operating patterns throughout the day;

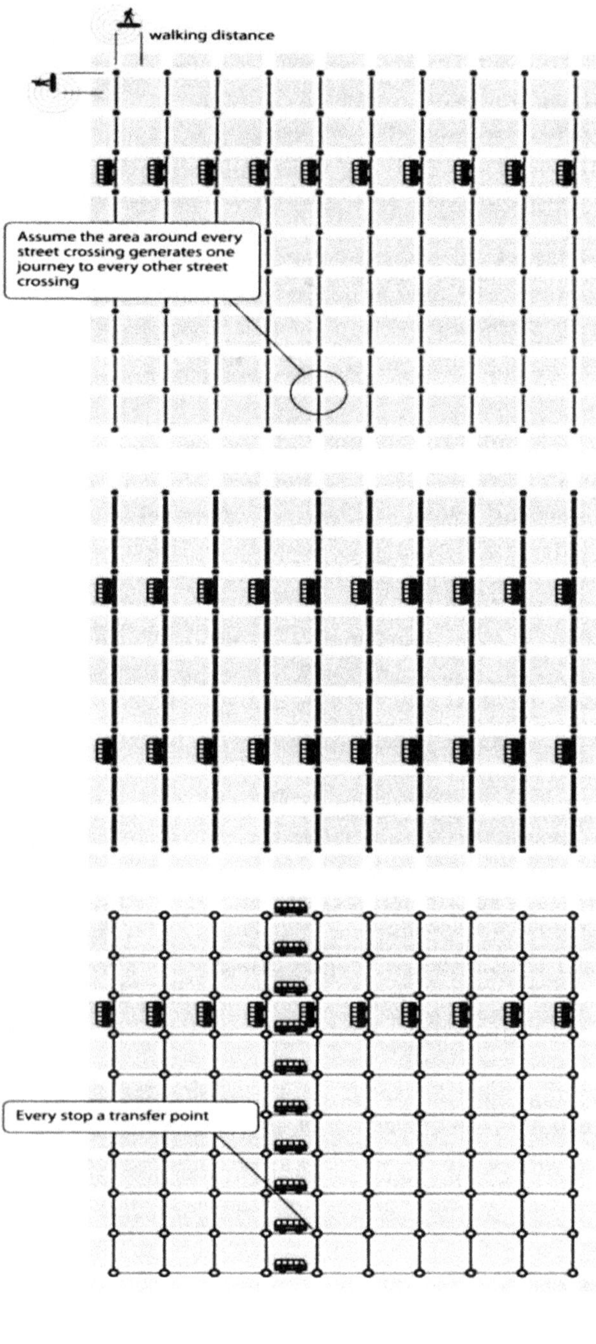

Figure 9.4 The network effect: the 'Squareville' example (reproduced, with permission, from *HiTrans Best Practice Guide No. 2: Public transport – planning the networks*, (Gustav Nielsen and Truls Lange, adapted from Mees, 2000: 138–42). See Box 9.2.

Box 9.2 The network effect, the Squareville example (see also Figure 9.4) (reproduced, with permission, from *HiTrans Best Practice Guide No. 2: Public transport – planning the networks*, (Gustav Nielsen and Truls Lange, adapted from Mees, 2000: 138–42)

1 **Squareville with ten bus lines running north–south**
The hypothetical city of 'Squareville' has a grid iron street pattern. The streets are well suited for a bus service since they are 800 metres apart. Squareville is a homogeneous city with a travel demand that is entirely dispersed. Assume that the area around each of the city's street crossings generates one journey to every other street crossing: 9,900 trips per day in total. For the whole of Squareville, the ten bus lines can only serve 900 trips in the city, which is less than 10 per cent of the total 9,900 trips. Assume that the public transport service presently attracts one-third of the journeys it can theoretically serve. This gives 300 trips per day by public transport, which is a modal share across the whole city of only 3 per cent.

2 **Squareville with ten bus lines running north–south, double frequencies on all lines**
Imagine that services on the existing bus lines are doubled in order to encourage more people in Squareville to use public transport. According to traditional transport demand modelling the elasticity of demand might be assumed to be some 0.5. This means that a 100 per cent increase in service will produce a 50 per cent increase in demand. The result will be 450 public transport trips per day and a modal share of 4.5 per cent. Since the operational costs are likely to increase by more than 50 per cent, the cost recovery through fares is likely to fall.

3 **Squareville with twenty bus lines running north–south and east–west**
Imagine that the extra operating resources, instead of doubling the existing service, were used to run ten new bus lines in the east-west direction. This would create a grid network of twenty lines. The number of trips that are directly served would double to 1,800: the 900 initial north–south journeys and the 900 new east–west journeys that can be made without transferring between lines. But if the passengers are willing to transfer, then all 9,900 trips between all blocks can be served by this network: 1,800 directly and 8,100 by transferring. Assume that the modal share for journeys involving a transfer is half of that for direct journeys, i.e. one sixth of those trips that can be attracted to public transport. This gives a total number of 1,950 public transport trips per day (one third of 1,800 plus one sixth of 8,100). The modal share has increased dramatically from 3 to 20 per cent. This gives an elasticity of demand that is 5.5, rather than the traditional figure of 0.5. Increased revenue from the fares should more than cover the extra costs of operation and the vehicle occupancy would rise.

- convenient transfers through coordinated timetabling and good physical design of interchange facilities;
- region-wide planning by a public agency and a centralized fare collection regime to allow cross-subsidies and free transfers. In this model, an appropriate role for the private sector is to tender for the right to operate services forming part of the network.

This conceptual work was reinforced with a detailed assessment of current public transport planning practice in the New Zealand cities. This assessment starkly demonstrated the inefficiencies created by many decades of ad hoc service planning. Together, the three components of the project led neatly to creative discussions of practical changes that could be made to local service planning and delivery.

The major political and institutional success of the project has been to build support among senior managers and other key players in local public transport planning policy networks for significant changes in service design and delivery. In Wellington, planners have developed two options for network improvement based on concepts generated during the GAMUT project. These have had a positive reception from key players, in part due to learning achieved during the project. In Christchurch, planners are using networking principles in the redesign of public transport services to be introduced as the city is reconstructed following the 2010 earthquake. Planners in Brisbane and Melbourne are taking some steps to leverage further patronage gain from investments, such as new high-frequency buses and cross-suburban routes, in ways that in part mirror the objectives of the project. However, without enhanced networking these investments may lead to sub-optimal results.

Full details of the project can be found in *Public Transport Network Planning: a guide to best practice in New Zealand cities* (Mees et al., 2010) and Stone, Mees and Imran (forthcoming).

How new analyses help resolve the dilemma of mobility

The tools and the applications discussed in this chapter present different ways of exploring key problems and arriving at solutions. Improved tools can overcome the limitations of previous approaches to exploring mobility in cities. They can improve the basic understandings of the components, linkages and relationships in transport systems. They can help improve analysis and assessment of key transport and land use problems. They can also help identify improved 'solutions' and policy or project responses. In particular, each of these practice technologies opens up solutions that do not depend on massive investment in new infrastructure, whether by changing land use arrangements and the location of new development, or in deploying existing services in a more efficient and effective manner. They allow policy makers and transport operators to optimize the use of the transport network, and obtain significant sustainability benefits.

There remain challenges. In particular, there are inherent tensions created by modelling, especially as models are full of choices, assumptions and the priorities of the modellers themselves, which are often not transparent. And stakeholders sometimes lack the ability or willingness to criticize, engage with or accept modelling results. Development of trust and increased stakeholder involvement remain critical goals and more research is clearly needed in this area. These issues are revisited in Chapter 11.

Notes

1 SNAMUTS has been developed in the context of several research and consultancy projects funded by public agencies in Victoria, Western Australia and the Australian federal governments since 2006 (Scheurer, Bergmaier and McPherson, 2006; Scheurer and Porta, 2006; Scheurer and Curtis, 2008; Curtis and Scheurer, 2009).

2 The preferred network path between a pair of nodes is defined by comparing travel times between the path with the lowest number of transfers and the path with the lowest cumulative impediment, including a realistic estimate of the time taken to do each transfer. The path with the lower resulting travel time is the preferred path.

10 Harnessing the energy of free range children

Carolyn Whitzman

Car dependence and its impact on children's independent mobility

Children are the transport users and decision makers of the future. The decline in children's autonomous exploration of public space has had disastrous consequences, for their health and also for community wellbeing. Conversely, taking a child's-eye view in making our communities safe, accessible, and welcoming may well improve outcomes for all. This chapter outlines a 'children's right to the city' approach to urban planning policies and practices that can encourage children's autonomous walking, cycling and public transport use in their neighbourhoods and cities.

This chapter takes a slightly different approach from that of the other chapters in this book. It focuses on one particular stream of the GAMUT research – 'institutional barriers and enablers to children's independent mobility' – and on one particular population group: children, particularly those aged 6 to 12, in primary school.

The 'automobility dilemma' discussed in the first chapter of this book has hit children particularly hard. The definition of children's independent mobility (CIM) is the ability of those under 18 to explore their communities autonomously, without adult supervision. This definition derives from foundational work in the UK by Hillman, Adams, and Whitelegg (1990). The lead author of that report has also used the terms 'battery reared' and 'free range' to refer to two diametrically opposed approaches to children's upbringing (Hillman, 2002). Particularly in wealthy English-speaking societies such as those of Australia, the UK and North America, children under 12 have seen their right to roam in public space radically eroded over the past two generations as a consequence of societal car dependency and a culture of fear. They have become 'bubble wrapped' (Malone, 2007) and 'ghettoized' (Matthews, 1995) by concerns about traffic safety and stranger danger. Children directly contribute to the abandonment of streets and public spaces through their absence. They indirectly contribute to car dependence through their parents chauffeuring them to home, play, and friends' houses, virtually trapped in their homes and back yards by a vicious circle: more cars leading to diminished safety for pedestrians.

As with adults and the journey to work, the most common measurement for children's independent mobility is the journey to school. The longitudinal research by Hillman et al. (1990: 42) found that in 1971, 80 per cent of 7–8-year-olds in the UK were allowed to travel to school on their own, mostly by walking, but in 1990, this had fallen to only 9 per cent. In the US in 1969, approximately half of children walked or bicycled to and from primary or secondary school, and 87 per cent of those living within one mile of school walked or bicycled. In 2005, fewer than 15 per cent of children and adolescents in the US used sustainable means of transport to get to or from school (Martin and Carlson, 2005: 2161).

Although the same kind of longitudinal data is not readily available for Australia, in New South Wales, Australia's most populous State, the proportion of children aged 5–9 years who walked to school declined from 57.7 per cent in 1971 to 25.5 per cent in 1999–2003, with a corresponding drop from 44 to 21.1 per cent in children aged 10–14 over that time period (Hidde et al., 2008: 60–62). In Melbourne, capital of Victoria – Australia's second largest State – and the focus of this research, 55.3 per cent of young people walked to school or higher education in 1970; this figure fell to 22.2 per cent in 1994. In the same time period, car travel to school increased from 14.3 to 43.9 per cent (Garrard, 2009: 9).

As is the case for adults and the journey to work, the journey to school represents only about one fifth of trips (Mackett, 2004). From 1994 to 1999, children aged 0–14 years in Melbourne made an average of 23.1 trips per week, 16.3 of which were as a car passenger (71 per cent), 5.0 as a pedestrian (22 per cent), and 0.2 as a cyclist (1 per cent). During this five-year period, Melbourne children spent on average four hours a week as a car passenger and 48 minutes walking (Ironmonger and Norman, 2007, unpaginated conference paper). When Karen Malone (2007: 516–17) asked 50 children in the regional Victorian city of Bendigo to take photographs of their typical week, over half included a picture of the back seat of their family car.

This combination of dependence on adult supervision and on cars has dire consequences for children's health. Much of the research on the health impacts of car dependence is carried out by public health researchers, who point out that sustainable transport is also active travel. That is, the energy required to move from one place to another can be accomplished through burning calories via walking and cycling or by burning fossil fuels (Hidde et al., 2008). Using public transport generally involves walking to and from a stop or station. Therefore, the agendas of health promoters and environmental activists should be aligned on the issue of active and sustainable transport.

In Australia, over one quarter of Australian children are overweight or obese, as compared to 5 per cent of Australian children in 1960 (Australian Society for the Study of Obesity, 2004). Simply put, obesity is the product of a 'chronic energy imbalance': people not exercising enough on a regular basis to work off the energy in the food they are eating (Baur, 2007: 4).

Health researchers have demonstrated that children who regularly walk or cycle to school and other destinations are more likely to meet the recommended daily minimum level of 60 minutes per day of moderate to vigorous physical activity than those who are driven, even if the latter children engage in organized physical education (Garrard, 2009, Mackett et al., 2004). Low levels of physical activity are associated with increased risk of heart disease, high blood pressure, type two diabetes, and colon cancer, through overweight and obesity, among children as young as 12 (Timperio, Salmon and Ball, 2004; Martin and Carlson, 2005). Public health researchers have blamed transport and land use planners for creating 'obesogenic environments' (Catford and Caterson, 2003: 578), where everyday physical activity is constrained and discouraged by built form and social rules such as those governing traffic speed – dangerously high.

There are also local environmental health risks associated with inactive and unsustainable travel to school and other destinations. Kingham and Ussher (2007: 54) cite a UK study that found the quality of air near schools can be worse than surrounding streets, partly because short car journeys associated with the school run are the most polluting. Tranter (2004: 11) cites the International Centre for Technology Assessment in saying that, 'in-car air pollution may pose one of the greatest threats to human health,' and the World Health Organization, who have found that the number of deaths caused by transport related pollution are greater than the number of deaths from vehicle accidents.

In contrast to the overwhelming health risks of inactive and unsustainable travel to school, are the relatively low, but perceptually important, safety risks of car traffic and stranger danger. Absolute risk of fatality for all modes of travel is low in most developed countries, including Australia, where the risk of a child being killed as a car passenger is twice that of dying as a pedestrian and more than four times that of being killed as a cyclist (Garrard, 2009: 4). Only 24 children were killed in pedestrian/car accidents in Australia in 2004, less than a quarter of those who were killed in 1972. Five children were killed in 2002, all by family members in their homes, as compared to 10 children in 1972 (Cadzow, 2004: 21). However, the decrease in pedestrian accidents may be related to children's declining use of public space (DiGuiseppi et al., 1997). A cluster of five countries – Sweden, the Netherlands, Finland, Germany, and Denmark – where relative risks of pedestrian/cyclist death and injuries are low, indicates that measures such as low urban speed limits, national road safety campaigns, and legislation that assumes driver responsibility in an accident involving a child pedestrian, have combined to improve pedestrian traffic safety. These countries have high rates of active transport for both children and adults, and low rates of obesity (Garrard, 2009: 6–7). In contrast, Anglo-American countries such as Australia place the onus on individual parents to protect their children, to the extent of criminalizing parents for letting their children walk and cycle (the case study in Box 10.1).

Box 10.1 An alternative to children's independent mobility: criminalize parents!

In April 2010, Raquel Nelson and her three children, aged 2, 4, and 9, went to a family celebration in the suburbs of Atlanta, Georgia. On the way home, they stopped at Walmart for groceries. The single mother had no car, so they took public transport.

On the way home, they just missed a bus transfer, and had to wait an hour for the next bus. When the bus stopped across the street from their apartment building, it was dark. They had to cross the four-lane road, which had no safe crossing point. The nearest 'legal' pedestrian crossing was 450 metres away. Together with several other parents and children who got off at this stop, they attempted to cross the busy road. At the median, 4-year-old AJ saw an adult attempting to cross the other two lanes and darted into traffic. His mother followed, attempting to keep him safe. AJ was struck by a van and killed, and his mother and 2-year-old sister were seriously injured.

AJ was killed by a drunk driver with previous DUI convictions along the same stretch of road. While manslaughter charges against the driver were dropped, Raquel Nelson was charged and convicted of vehicular homicide in the second degree (Goodyear, 2011). This is hardly the first time that low-income single mothers without cars have faced criminal conviction when their children were killed by speeding motorists – there was a similar case in the same city just one year previously, involving the death of another 4-year-old child (Peds, 2009). The Atlanta metropolitan region ranks 11th in the US for most dangerous roads, and had nearly 800 pedestrian deaths between 2000 and 2009 (Snyder, 2011). Motorist or government responsibility for this high death rate appears not to have been considered.

In 2005, a UK car insurance company sued the parents and childcare provider of an 11-year-old cyclist severely injured by a car driver for not ensuring she was wearing her helmet at the time of the accident (Gill, 2005: 40). More recently in London, Oliver and Gillian Schonrock were told by the principal of their children's school that they would be reported to Child Protection Services if they continued to allow their 8- and 5-year-old children to bicycle together unaccompanied by an adult, although the distance from their home to school is less than one kilometre. Ironically, Cycling England has recently released a report criticizing 'overprotective' parents for stifling their children's desires to cycle to school (McVeigh, 2010).

Declining physical health is only one of the negative health impacts of unsustainable transport. Equally important are the threats to social and emotional wellbeing. Prezza et al. (2005: 437) summarize the impacts of reduced autonomous walking and exploration of the local environment: reduced environmental knowledge, retarded development of social, motor and analytic skills, and fewer local friends and acquaintances. Kingham and Ussher (2007: 503–4) cite evidence of how regular exercise can enhance children's psychological wellbeing by improving moods and reducing anxiety. Formal playgrounds such as sports fields and school yards are no substitute for the right to explore entire communities. Indeed, they can be seen as 'a paranoid attempt to create "no risk" environments for children' (Hart, 2002: 135), with the number of formal playgrounds being inversely related to children's freedoms (Ward, 1978).

The United Nations Convention on the Rights of the Child has been ratified by the majority of countries in the world, including Australia (UNICEF, 2008). As will be discussed below, the Convention addresses children's rights to participate in their communities and also to autonomous use of public space. Engwicht (1992: 39) eloquently summarizes the impacts of car dependence on children's citizenship:

> Changing streets from places to 'movement corridors' robs children of the opportunity to explore their neighbourhood in ever increasing circles as they mature. This freedom to explore the local neighbourhood is probably the key ingredient in children feeling they belong to a neighbourhood, a place. It not only gives them an opportunity to develop relationships with people of all ages who live in their neighbourhood, it gives them the opportunity to develop a relationship with the *placeness* of their physical environment. Robbing children of a sense of place robs them of the very essence of life.

In contrast to the emphasis on the negative impacts of car dependence on children, a growing number of researchers point to the possibilities that might be unleashed by listening to children's views on their local environments. Tranter and Sharpe (2008) use the metaphor of the children's movie *Monsters Inc.* to describe the emancipatory potential of planning cities for the free and active mobility of children. While 'child *un*friendly cities are premised on the notion of cheap, available oil, where consumerism and private car transport increase the dependency of children upon their parents and other adults, and effectively place much of the city "off limits" to children' (Tranter and Sharpe, 2008: 295, emphasis in original), the dual challenges of climate change and peak oil '[provide] opportunities for societies to re-conceptualize children and their role in society' (ibid: 298). Perhaps the 'strategy of engendering laughter by allowing children to play freely and safely in their cities represents an enduring way forwards' (ibid: 306) to a more sustainable society for people of all ages.

Which policies promote children's independent mobility?

The aims of the four-year GAMUT project were to identify policy interventions that facilitate children's independent mobility within urban environments, and to determine whether there are principles across interventions that can be replicated in different settings. The emphasis was on the role of both land use and social planning in achieving higher levels of children's 'right to the city' and thereby improving health, citizenship, and sustainable transportation outcomes. While the geographic focus was Victoria,

Australia, we drew on a wide range of Australian and international research and policy literature, particularly in the search for best policies and practices.

Our research had two phases. In the first phase (July 2006–December 2007), our objectives were to:

1 overview the evidence base for making children's independent mobility a policy priority within the State of Victoria, Australia;
2 accomplish a first scan of research on what works and what might work in an Australian context to enable CIM; and
3 identify key stakeholders, policies and programmes within Victoria.

We utilized three main methods in this first phase. First, we undertook a review of Australian and international literature related to CIM to determine the scope of the problem, and to identify factors that acted as barriers and enablers. Second, we undertook a content analysis of 20 key Victorian policies and practices, to determine how and to what extent they directly or indirectly address CIM. Third, in addition to dozens of ongoing discussions with researchers and practitioners, we interviewed 14 Victorian decision makers, researchers, and advocates on the issue of CIM. The results of the first phase of research are summarized in the report *From Battery-Reared to Free Range Children* (Whitzman and Pike, 2007).

An early finding was that the causes of declines in CIM were complex and inter-related. A conceptual framework explaining a primary school child's travel behaviour might look like Figure 10.1.

In this model, urban planners and transportation engineers might focus on the urban form determinants of adult-dependent mobility, such as increasing traffic volumes and speeds, increasing sprawl leading to

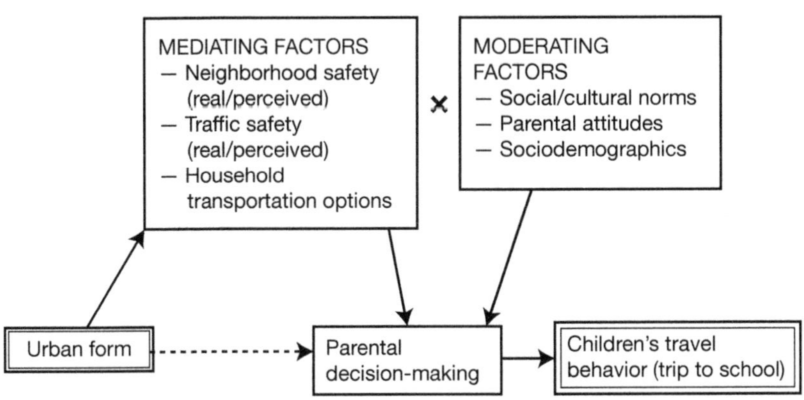

Figure 10.1 Conceptual framework of a primary school child's travel behaviour (McMillan, 2005: 449)

fewer nearby travel destinations such as schools, shops, and play spaces, decreasing public transport service, and other aspects of urban form. But these material realities are moderated by perceptions of neighbourhood safety, parental norms – no one wants to be the first parent in a community to let their child walk or cycle to school – and whether there is a critical mass of children and adults walking and cycling in the community. These factors combine to determine parent–child negotiation of 'licenses': the permission granted by parents for children to travel to school alone or with friends, play in the street or in a nearby park without adult supervision, travel to other places in their local environment, take public transport alone, cycle on main roads, and go out after dark (Hillman et al., 1990).

Another explanatory model, this time from the field of public health, brings in the policy environment (Figure 10.2).

Policies and practices can determine social and cultural norms, parental negotiations, and the physical environment in ways that can influence travel behaviour either negatively or positively. In our survey of the international literature, we found four categories of policies and programmes that have been hypothesized to influence CIM.

First, high levels of traffic volume and speed increase pedestrian risk and decrease use and enjoyment of streets (Garrard, 2008). For this reason, planners and designers have promoted 'traffic calming' measures such as reduced speeds and road widths, and in some cases, closing off or severely limiting the use of motorized vehicles on residential or commercial streets (Prezza et al., 2005; Engwicht, 1992). Gill (2007) argues that low-income regeneration areas should be priorities for traffic calming, along with new development and those areas already showing strong resident support. Both Gill (2007) and Prezza et al. (2005) support additional measures, such as benches, greenery, children's mosaics, and other public art that may support children's travel and play on streets. There is strong evidence that reduced vehicle speed leads directly to increased active travel behaviour among both

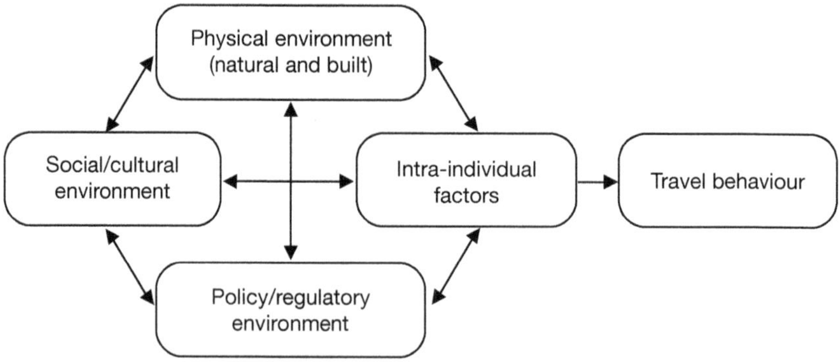

Figure 10.2 Environmental influences on children's active travel behaviour (Garrard, 2009: 11)

adults and children, as well as having indirect impacts, through improved perceptions of traffic safety, community amenity and reduced risk of injury (Garrard, 2008).

Second, the expanding land use division of the twentieth century, with residential areas separated from employment and shopping districts, has been hypothesized as both a cause and an effect of urban sprawl and resultant car dependence (Newman and Kenworthy, 1999). Planning and design that is supportive of both walking and use of public transport promotes access to nearby destinations, through increasing population densities that can support schools, shops, and recreation facilities in communities. Unlike the strong and proven relationship between traffic calming and increased use of active travel, there are few studies that demonstrate – for example, through before/after intervention/control experiments – that these measures increase active travel in general, and CIM in particular.

Third, increasing fears of car traffic and stranger danger are the most commonly cited reasons among both parents and children for abandonment of public spaces and streets (Prezza et al., 2005; Timperio, Crawford, Telford and Salmon, 2004; Hillman et al., 1990). A number of programmes have been undertaken that combine social marketing with design changes, such as *Safe Routes to School* (McMillan, 2005). *Safe Routes to School* involves education of both children and drivers on road safety, enforcement of traffic laws around schools and re-engineering the streets along the routes to schools. However, the evaluation of these initiatives has shown mixed success (Boarnet et al., 2005; Timperio, Salmon and Ball, 2004). Another school-based programme, the *Walking School Bus*, supports active travel by getting parents to volunteer to shepherd students to school along a set route. While evaluations have shown some improvements to active travel and social cohesion outcomes, the programme appears to support adult-dependent mobility, and has shown, at best, equivocal results in terms of CIM (Kingham and Ussher, 2007; Kearns et al., 2003).

Finally, there is the argument that modern planning and design has created an 'archipelago of "safe spaces" [for children] in a sea of adult-centric space' (Fincher and Iveson, 2008: 107). In contrast is a rights-based approach exemplified in *Child-Friendly Cities*. This international programme is sponsored by UNICEF in support of the Convention on Children's Rights. It supports local governance efforts to fulfil children's rights to influence decisions about their city, participate in family, community and social life, and walk safely on the streets on their own (UNICEF, 2008). In other words, it supports children's full integration into public space and public life. While the most clearly supportive programme in terms of CIM, *Child-Friendly Cities* is a process, rather than a standardized set of initiatives. As such, the programmes have lacked rigour in terms of evaluation, particularly in the area of increased CIM (Bartlett et al., 1999).

Turning to the Victorian case study, despite the decline in CIM being seen as a concern by the policy makers and researchers we interviewed, the first phase of research found no Victorian or national policies that explicitly targeted this issue (Whitzman and Pike, 2007: 29–39). At this point, it should be noted that during the period of the initial research, there was no national policy explicitly connected to urban planning, and the national level of governance is still not a major player in urban policy. As noted in Chapter 7, the State government is responsible for planning legislation, major roads and public transport policy, and major health policy. Local governments are responsible for local roads, local planning decisions, and local health promotion.

Overarching State government policies, such as *Growing Victoria Together* in 2001, which sets out government priorities and is regularly updated for progress on indicators, did not address CIM in the context of 'sustainable transport', 'protecting the environment for future generations', or 'building cohesive communities and reducing inequalities'. In 2004, *A Fairer Victoria* endorsed a 'life cycle approach' to reducing inequalities and improving social sustainability, but focused on direct service provision to at-risk children rather than more upstream and diffuse health promotion efforts aimed at improving CIM. *Melbourne 2030*, the metropolitan strategic land use plan intended to guide urban growth, had little explicit consideration of the needs or opinions of children. *Neighbourhood Renewal*, which guides efforts to narrow the gap between the most disadvantaged neighbourhoods in Victoria and the rest of the State, likewise had little explicit consideration of children within these neighbourhoods (Whitzman and Pike, 2007: 29–39).

However, a number of initiatives were identified as promising. *Go For Your Life*, a whole of government initiative that sought to improve healthy weight outcomes, was focused on children's healthy eating and organized sport, rather than active travel. This initiative was discontinued by the State government in 2008. *Walking School Bus* initiatives were financially supported in the majority of Victoria's 79 local governments by the Victorian Health Promotion Foundation (VicHealth) at the time of the initial research, although this programme was also soon de-funded. *TravelSmart*, a joint State and federal social marketing initiative, targets schools, workplaces, and the wider community. It engages children and adults in analysing their current travel behaviour and improving the modal share of active transport. Although the evaluations are weak – as they focus on short-term self-reported behaviour change – it shows potential and is still continuing. There were a number of bicycle education programmes operating in schools at the time of the initial research, and also an initiative supporting public transport use by older children called *Public Transport Challenge* – which no longer exists. There is also a *Safe Routes to School* programme that, in contrast to the social marketing approach of most initiatives described, supports changes to the physical environment around schools and provides road safety information to children and parents. Unfortunately, the travel safety messages often include messages unhelpful to the notion of gradually increasing CIM, such as the injunction that an adult always hold the hand of a child under 11 (Whitzman and Pike, 2007: 29–39, 11).

The most promising initiative identified in the initial research was *Child-Friendly Cities*. As discussed in the introduction, its principles include a citizenship approach that explicitly includes children's right to influence decisions about their city, walk safely on the streets on their own, and be an equal citizen of their city, with access to every service (UNICEF, 2008). At the time of the initial report, the City of Greater Bendigo, a regional Victorian city with 86,000 inhabitants, had become the first Australian city to apply for formal accreditation as a *Child-Friendly City* – it received accreditation in 2007. Several other local governments, including Port Phillip and Melbourne in the inner part of metropolitan Melbourne, and Brimbank in the west of metropolitan Melbourne, were also identified by interview participants as pursuing *Child-Friendly City* initiatives.

Child-friendly cities as an enabler of children's independent mobility

By the end of 2007, it was clear that a 'window of opportunity', as it is labelled elsewhere in this book, had opened up in Victoria, and possibly across Australia and New Zealand. In the case of children's independent mobility, the window of opportunity consisted of a critical mass of locally based research, from both public health and planning disciplines, showing a serious challenge to children's health. This, in turn, led to action from local politicians and staff, and State-wide advocacy agencies. Finally, there was some level of institutionalization through the *Victorian Child Friendly Cities Network*, albeit one that was limited in power, since it was based in the local rather than the State level of governance.

The changes were led by three non-profit advocacy organizations. The Victorian Health Promotion Foundation (VicHealth) was initially funded through a State government tax on cigarettes, but its mandate has grown over the past 20 years to encompass all aspects of health promotion, including promoting physical activity. In late 2007, VicHealth decided that its support of the *Walking School Bus* was not achieving sufficient results in terms of increases in children's physical activity, so they shifted their funding priorities to a more flexible and targeted intervention in six localities. The new programme, called *Streets Ahead*:

> Builds on the learnings from VicHealth's *Walking School Bus* program, which encouraged children to walk to and from school under adult supervision. *Streets Ahead* goes further. The idea is that for short local trips we can help children to use their own legs and wheels more – and travel by car less – by encouraging them to *walk, cycle, skate or run* to school and other local destinations. It's about using active transport instead of being a passenger and, where age appropriate, without adult supervision.
>
> VicHealth, 2008: 1, emphasis in original

The programme, which works with partnerships of schools, local governments, community organizations and parents – but without children in an explicit advisory capacity – was independently evaluated in late 2011, after three years of operation.

At the time of our initial report, *Bicycle Education* had been delivered by the Police in partnership with VicRoads, the State government organization that plans and manages roads – in contrast to the Department of Transport, which plans and manages public transport, walking and cycling. There had also been a pilot project called *Cycling for a Sustainable Future* sponsored by Environment Victoria, an advocacy organization, with a more explicit commitment to overcoming barriers to children cycling independently (Whitzman and Pike, 2007). As of 2007, Bicycle Victoria took on the major role in providing *Bicycle Education* in schools. It organized a National Bike to School day in 2007, which has since been repeated annually. It has also shown a considerable commitment to evaluation of its programmes, which have grown considerably in terms of numbers of schools, numbers of children participating, and increased number of children using active travel to school (Garrard et al., 2009).

The Heart Foundation, like VicHealth, had undertaken a campaign on supportive environments for physical activity as part of its health promotion mandate. VicHealth and the Heart Foundation supported a forum, under the aegis of *Go For Your Life*, on 'Reshaping How We Live: the Influence of the Built Environment on Health' in April 2008, and another *Go For Your Life* forum targeted at 'Children and Active Transport' in October 2009. The Heart Foundation is currently supporting a programme called *Space for Active Play*, in which children aged 8–12 in several Victorian communities are redesigning parks to encourage their active and independent use. Evaluation of this programme in terms of increased children's use of parks is under way.

The State government has been less successful in developing new policies and practices that support CIM and active travel, although several significant changes have been implemented that may deliver better active travel outcomes for all citizens. VicHealth and the Heart Foundation, again working to complement one another's programmes, advocated changes to the State-wide residential planning provisions, leading to 'sustainable neighbourhood' policies that promote walking and cycling (City of Greater Geelong and David Lock Associates, 2009). They also lobbied for changes to the Planning and Environment Act, the State legislation that governs local planning, that would enshrine social and health considerations in new land use planning decisions (National Heart Foundation of Australia, 2009). VicHealth worked with the Victorian chapter of the Planning Institute of Australia to develop a *Planning for Health and Wellbeing* project that educated land use planners on the need to plan health into communities, while the Heart Foundation's Victoria-based *Healthy by Design* guidelines (2004) were followed by a national project on *Healthy Spaces and Places* (Moroney, 2009), with the national office of the Planning Institute of Australia and the Australian Local Governance Association. Despite these significant gains in terms of advocacy and partnerships, the State government's support for car dependence through funding priorities, planning policies, and research, has proven to be a weighty 'supertanker' to turn around.

Local governments in Victoria have proven more innovative than state government. By the time of the first report of this project in 2007, the City of Greater Bendigo had consultations in progress with 500 children on what makes their community a good place for them, and what would make it a better place. This included an art competition, with 86 entries, that invited children to tell the local government what makes their street child-friendly. Answers included: being able to play, ride a bike or scooter, kick a ball or play cricket on the street; having friendly neighbours who waved or talked to them; and having cars that go slow (Whitzman and Pike, 2007: 38). As these consultations appeared to be in the process of being translated into policies supportive of CIM, *Child-Friendly Cities* became the focus of the next phase of research, which was undertaken from January 2008 to July 2009.

In the second phase of the research, we asked the question: 'Can the development of child-friendly cities promote children's independent mobility?' Our emphasis this time was on current land use and social planning policies and practices. We began by focusing on five local governments that we knew were pursuing *Child-Friendly Cities* initiatives: the Cities – a term used to describe urban local governments – of Melbourne and Port Phillip in the central city, Brimbank as a middle suburb, and Bendigo and Ballarat, two regional cities. We matched these with two 'control sites' with no apparent interest in *Child-Friendly Cities*, one in the central city and one in a middle suburb – all regional cities surveyed were undertaking *Child-Friendly City* initiatives.

Based on the literature, we developed a model with six elements of ideal policies and practices that might promote CIM. First, the effectiveness of planning policies is dependent on the explicit recognition of children as an interest group, thus rejecting the 'trickle-down' approach that assumes that what is good for adults is good for children (Bartlett, 2005: 19). Planning for the 'universal child' is insufficient, as well, because children are diverse in terms of income, ethnicity and disabilities. Second, planning policies must explicitly recognize children's rights to all public space, not only those designated as 'child-specific' spaces such as playgrounds or schools. Third, policies must provide achievable targets, strategies and implementation mechanisms that name lead departments. These indicators might include: an increased number of children reporting use and comfort in public spaces, whether there are basic services like shops, schools, and recreation areas within easy walking distance, and whether children report that these destinations are accessible and pleasant to use. Fourth, policies that consult with and support children must be integrated into other local government policies, such as health and land use plans (Whitzman et al., 2009: 19–21).

There were two further criteria for model practices. First, social and land use planners, local councillors and senior managers must be trained in a rights-based approach to planning for children. It is not enough to train one policy champion, since turnover is high in local government. Second, planners must be equipped with the skills that allow them to interact with children, including the ability to transform complex plans into simple diagrams, and complex ideas from children into planning policies.

We used this model to evaluate policies and practices in all seven local governments. We found that those local governments committed to child-friendly cities were more likely explicitly to recognize children as a group with rights. For instance, the City of Port Phillip was committed to developing a child impact assessment tool for policies and services, and also committed to including in council report templates a requirement to state the effects of any proposal on children. In contrast, the major health and land use planning documents in the two control sites emphasized 'families', not 'children', as a group with specific needs (City of Port Phillip, 2005).

We also found that local governments committed to child-friendly cities had commenced integrating children's right to walk and explore in cities. The City of Port Phillip has embarked on a *Green Light* campaign to lower speed limits and increase traffic stopping times, particularly around schools, while the City of Melbourne has recognized, in its recent long-term strategy plan, that 'whether or not children are safe to walk is a real test of a walking city' (City of Melbourne, 2009: policy 3.1). In contrast, the control sites were still focused on direct services such as child immunization and the management of specific spaces like child care centres and playgrounds.

We found that both control and intervention sites were weak on specific targets and monitoring systems. The City of Port Phillip's *Child-Friendly Cities* strategy did speak of key progress indicators, but they were generic goals such as 'nutrition and weight measures' and 'safe and accessible environments' (2005: 2). In general, the listing of resources, lead council business units and timelines was underdeveloped in relation to policies directly affecting children (Whitzman et al., 2009: 28–9).

All of the local governments committed to child-friendly cities were in the process of institutionalizing participation with children. For instance, the City of Ballarat, in its initial report on becoming a *Child-Friendly City*, recognized that children constituted an underrepresented group in public consultation relative to their demographic weight, and that 'further views need to be sought from our Under 19 years' (City of Ballarat, 2007: 9–11). However, the land use planning strategies in both intervention and control sites did not mention consultation with children in relation to their Municipal Strategic Statement, which is their overarching land use planning policy (Whitzman et al., 2009: 25–6). In general, land use planning policies were still poorly integrated with health and social plans (ibid.: 2009: 26–8).

In each of the seven local governments, we interviewed one council officer directly involved in either a *Child-Friendly Cities* initiative – in the case of the five intervention sites – or in children's services – in the two control sites – and one land use planner. We found that in all of the intervention sites, land use planners have a more comprehensive understanding of children's rights and the importance of consultation. However, most land use planners reported that, in practice, children are rarely consulted in relation to neighbourhood land use strategies. Usually, consulting with parents is considered an effective substitute (Whitzman et al., 2009: 30–1). Bendigo and Brimbank were involved in *Streets Ahead* initiatives funded by VicHealth, although at the time of the interviews in March–April 2009, neither had yet consulted with children on active travel (ibid.: 21–32).

The local governments we examined varied widely in terms of the training they had undertaken. The Cities of Port Phillip and Greater Bendigo were furthest ahead in terms of training local politicians and senior staff, as well as social and health planners. There was the sense among all local government staff we interviewed that training for land use planners, as well as local politicians and senior staff, was essential for a whole of government approach, and that the child-friendly cities were still in their infancy in terms of a comprehensive approach to practice.

Conclusion: children's right to the city

In November of 2009, a Victorian *Child-Friendly Cities* Network was established, with an inaugural meeting that included 50 people from 22 local governments – the author spoke at that meeting. All of these local government council officers, who were mostly from social and health planning departments, expressed enthusiasm about consulting with children as a way of enabling their further participation in public space and public life.

As described in this chapter, this is a promising institutional enabler to children's independent mobility. Another enabler identified in this project has been a critical mass of State-level advocacy organizations – particularly VicHealth and the Heart Foundation – willing to fund both research and policy interventions to support children's active travel and autonomous exploration of public space. These State-level advocacy organizations, armed with good evidence, have been effective in lobbying the state government for changes to legislation, such as the Planning and Environment Act and residential development planning provisions – the regulatory framework – that would support active travel by walking, cycling, and using public transport.

There are still considerable barriers to overcome in order to have a societal impact on the precipitous decline in children's active travel and access to public space. One institutional barrier is the 'silo mentality' that still pervades governance, in Victoria and elsewhere. At the State level, the quasi-independent organization responsible for major roads in Victoria, VicRoads, is distinct from the Department of Transport, which is responsible for policy in relation to walking, cycling and public transport. In fact, the two agencies report to different ministers. Until 2008, the department responsible for community development was separate from the department responsible for land use planning. Both are still separate from the department responsible for urban regeneration in low-income neighbourhoods, which in turn is separate from the department responsible for public health promotion. In this light, whole of government approaches such as *Go For Your Life* can be forgiven for focusing on social marketing campaigns targeted at individuals and families. It is just too complex to address the institutional barriers to active and sustainable transport, such as a wholesale shift from primarily funding roads for the use of the private motor vehicle to prioritizing walking, cycling and public transport, or an end to urban sprawl.

Another barrier is the unwillingness of government to take seriously the notion of children's 'right to the city'. As described in Figure 10.3, a children's rights approach would emphasize the capacity of children to describe likes and dislikes in local environments, and their ability to recommend changes. Policies, programmes and practices would move beyond the provision of playgrounds for younger children and recreation centres for older children, to reinforce children's rights to explore autonomously throughout public space. This would go beyond the journey to school to encompass access to shops, public open space, friends' houses, and other destinations. Evaluation of initiatives would be based on increasing active travel, including exploration of public space without adult supervision. Land use and social planners would have both capacity and willingness to consult with children and translate ideas into concrete policies.

It is also true that there are large barriers to be overcome in parental perceptions. Even if car traffic were slowed down and limited, and even if there were easier access to schools, play spaces, shops and other local destinations, there would still be the issue of perceived risks to children in relation to traffic safety and stranger danger. In this research, we identified programmes – *Streets Ahead*, *Safe Routes to School*, *Green Light* – that

Safety through parental dependence	**Safety through children's autonomy**
Risks of traffic safety and stranger abduction highlighted	Physical health and social wellbeing potential of children's autonomous play highlighted
Parents are to blame if something bad happens to the child, including blame for children's inactivity	Emphasis on responsibility of government and civil society for children's wellbeing and safety
Role of experts is to tell children and parents what to do	Role of experts is to listen to children's ideas about local environments, and act on recommendations

Figure 10.3 From battery reared to free range children: a children's rights approach

sought to address both built environment and social perception factors simultaneously. It is possible that these programmes, in conjunction with local government initiatives like *Child-Friendly Cities* and policies that reflect this cultural shift, might be able slowly to reclaim the concept of children having a right to independent exploration of public space.

The movement to gather children's opinions and canvass their needs in relation to land use planning is still in its infancy. *Child-Friendly City* initiatives are still housed within local government family service departments, engage with social and health planning policies, and are not yet championed by land use planning staff. While offering a promising practice in terms of both physical and social environment change at the local level, *Child-Friendly Cities* initiatives must become more powerful within local governments as advocates for children as 'current citizens' rather than 'future citizens' (Simpson, 1997: 908–9). Until children's rights to travel autonomously within, and explore public space are entrenched in land use planning policies and practices, the full power of this approach to energize progress towards sustainable transport futures will not be realized.

11 Disseminating learning

Marcus Wigan and Crystal Legacy

Introduction

Central to achieving sustainable transport are the processes of dissemination of learning and engagement with the public necessary to generate the motive force to transform transport. Dissemination has usually been considered to be a task undertaken at the end of a project, and to be one where materials, usually printed, are made available some time after the completion of the projects involved (Box 11.1). This model of dissemination was deemed to be insufficient for the GAMUT initiative, which included both the classical dissemination methods and a more active and engaging process contemporaneous with research, with fewer barriers to accessing the broader knowledge gained. This chapter addresses the theoretical and practical aspects of dissemination processes as an intrinsic part of the development and execution of the GAMUT project.

Box 11.1 Aspects of dissemination

Dissemination of learning has a number of aspects, and in total they amount to more than the sum of the parts. Examining each in the context of the GAMUT project provides a rounded view of the knowledge, skills and social capital gained and how it can continue to be deployed. The categories identified here are:

- *Technical*: papers, books, journal articles, reports and websites. A range of GAMUT Working Papers has been produced, and many conference papers were delivered and journal articles published.[1] A special issue of *Australian Transport Research* (March 2011) was produced based on revised papers initially presented at the GAMUT international conference in 2010, with the objective of communicating some of the links between engineering and planning approaches to transport.
- *Personal*: participation in conferences, seminars, presentations to professional forums and visits. GAMUT's sponsoring body, the Volvo Research and Educational Foundations (VREF), has an active programme to

ensure communication and dissemination between their Centres of Excellence, with annual meetings of all Centres at different locations, and with special financial provision to allow the Centres' PhD students to attend and participate. A strong collaboration was established between the GAMUT Centre in Melbourne and the OMEGA Centre at University College London, resulting in a continuing theoretical and methodological dialogue around the research into transport mega projects – with GAMUT contributing three case studies and a successful PhD dissertation. The dissemination of GAMUT's ideas and research work beyond this VREF community has also been done through the conventional means, with papers delivered at the Transportation Research Board Annual Meetings, The Australian Transport Research Forum, the International Association of Travel Behaviour Research, the NECTAR group working meetings in Europe, meetings of the Association of European Schools of Planning (AESOP), World Planning Schools Congress (Perth, Australia, 2011) World Academy of Science, Engineering and Technology (WASET, Paris, 2010), State of Australian Cities conference (SOAC), and many other international and national forums.

- *Expertise*: education of graduate students, and the knowledge capital engendered and encapsulated by the participants for further exploitation. One of the key outputs from GAMUT has been the human capital generated. Four PhDs have already been awarded, with the successful graduates teaching and undertaking research elsewhere. Delivery of a Masters subject (Sustainable Transport and Public Policy) has equipped future professionals in environmental, urban and transport planning with the capacity to engage with the sustainability agenda. The human capital gains are not restricted to the graduates, as the interests and experience of all GAMUT participants have been expanded and changed by their experience.
- *Organisational*: forums designed to expand the platform on which further communication and learning can take place, a category that includes the social capital generated thereby; and
- *Research*: exploration of new ways of transferring both process and content into policy and practice. Research outputs reiterate the first four categories.

Disseminating learning as an exchange of knowledge

Transport planning has traditionally been mainly about technological fixes and finding solutions to narrowly conceived problems. With a few notable exceptions (for example, Dudley and Richardson, 2000; Vigar, 2002; Rietfeldt and Stough, 2007; Curtis and Low, 2012), the political-institutional aspects of transport planning, broadly conceived under the rubric of 'governance', have received much less attention. Drawing on the deliberative democracy and collaborative planning literature, we take a theoretical approach to conceptualizing processes of urban change towards sustainability, and in particular the knowledge exchange that takes place as part of these processes. The focus is on the engagement of stakeholders, including the lay community, in transport planning, and the interface between the policy makers, the political tiers of government, the users – for example, the community – and interest groups.

The growing subset of expert members of the community is explored here as a critical fresh component of this engagement. By this we mean the members of the community with high levels of expertise in, or related to, planning and transport. This group has been growing as governments divest themselves of internal expertise in these areas, and outsource them to consultants. However the problem for individuals – within or outside government – is the unclear and ever shifting boundaries of conflicts of interest, and the resulting disempowerment that can result from professionals in the public service, or consultants retained by government, being unable to operate in their private capacities to enable substantive community-based engagement with governments that they advise or serve.

A different democratic process

Planning theory has witnessed a transition from rational-based planning whereby the technical and expert knowledge possessed by policy makers frames substantive planning problems – i.e. alleviation of congestion, mitigation of and adaptation to climate change, management of population growth – to deliberative forms of planning that are inclusive, and where power imbalances are neutralized through careful process design. In particular, ill conceived – or at least incomplete – policy responses to substantive planning challenges have lacked the rigour that inclusive engagement of stakeholders offers to policy making. Policy emerging from traditional top-down processes now moves into a public domain with a set of stakeholder groups that increasingly regards these policy decisions as having been developed with no real attention to community ownership. When the community and its experts have been excluded from the process, such policies are thus seen as of questionable legitimacy.

Another concern developing as a result of maintaining the more conventional approaches to policy making is lack of transparency. In this case the rationale behind decisions is broadly unknown, and the critical examination of possible policy avenues is far from clear or explicit, and may even be regarded – sometimes correctly! – as having been managed to reduce proper appraisal of alternatives of interest to specific end user or impacted communities. Partly in response to this form of community disempowerment, deliberative democracy has evolved as an inclusive (Smith, 2003) and unconstrained (Dryzek, 1990) process of engagement that produces policy outcomes that are arguably more rational (Dryzek, 1987, 1990; Smith, 2001) and legitimate (Selman, 1999) than conventional democratic models, which in countries like Australia, the United States and Canada include varieties of pluralism, corporatism and neo-liberalism. These conventional democratic models are limited by highly unequal power relations among competing interest groups, and by their support for individualism and the pursuit of private gain. These features contradict any commitment governments may make to resolve collective societal challenges collaboratively (Held, 1987; Martell, 1994: 141).

The role of information – especially *transformed* information such as analytical and modelling outputs – is a key and growing factor. Just as the emergence of eDemocracy as a viable mode of policy development and engagement depended on both identified and anonymous communication processes enabled by recent ICT technology (OECD Expert Group on Government Relations with Citizens and Civil Society, 2003), so too there is a clear potential to expand the engagement processes where analytical and data-rich resources are involved. Evaluations of either eDemocracy or the enablement of better information bases for exchange are still scarce (Henderson and Hogarth, 2007).

The arguments for deliberative processes are well developed, and beginning to be assessed for their potential improvements in efficiency in the policy process (for example, Nabatchi, 2010). They are based on the capacity to encompass a broader spread of values than is normally included in government processes. The arguments bear repeating here as a basis for the modes of operation that are embedded in these perspectives. There are two apparently radically different approaches and philosophies involved:

1 What is often termed 'rational planning', where expert knowledge and skills are marshalled by the policy process to address emergent problems; the framing of the subsequent decision debates being entirely conditioned by this approach, leaving the community and diverse value engagement processes a narrow margin within which to try to moderate or redirect the momentum of policies.
2 A whole of community deliberative process, which includes a wider range of values and objectives, encompassing both political and idealistic goals and values held within the communities of experts and general populations alike.

Clearly the first would appear to its adherents to be 'efficient', 'decisive', and 'well targeted', while the second would seem to be inclusive, unmanageable and creating substantial overheads in dealing with complex issues where many interests have to be balanced as well as arguments underpinned by technical analysis. Many on the rational side might argue that community deliberation was taken care of by the political process of electing governments with particular mixes of values and objectives. However, in practice a mix of these approaches is commonly used, with efforts to change the policy agenda being very far from unspoken or absent in the negotiation – or otherwise – of final executive decisions.

A major common factor underpinning this apparent disjunction is the negotiation of information, data, and analytical processes and teasing out of consequences and impacts of the many decision variables. This negotiation process is crucially dependent on the degree to which the policy formation, framing and decision processes can function effectively. In turn, these depend on both the availability of the information basis, and the transparency of the analytical processes – if any – that are used to transform the information into a specific – and usually restricted – set of options.

The deliberative process has a long history of development, and is discussed next. But the very rapid rise in community expertise, education and capacity to handle information and engage in informed and demanding policy development and option assessment is frequently underestimated. We will subsequently address the consequences of these developments, and the major contribution that contestable evidence-based policy processes, which can make productive use of broader access to information, are now well positioned to make.

A 'whole of community' deliberative process

The rational hierarchical approach has benefits at the political level, as it permits political and bureaucratic control of evaluation, recommendations on rational and defensible premises, and decisions, and thus allows bureaucrats to manage the debate and elected officials to decide the outcomes. Late stage community distress, and the restricted scope for community engagement at an early stage that this model provides, is, of course, a design feature of the process. It allows the debate to be framed within restrictive and manageable boundaries, and can only have two outcomes: a decision for no action, or a selection from the limited range of options presented. The power of the information and analysis-based policy support thus carries a major momentum of credibility – until the community is capable of challenging it on its own grounds. Legal cases in the USA and UK challenging the analytical tools themselves have begun to appear, adding to the legal complexity of planning appeals and processes.

The evolution of deliberative democracy has coincided with increasing interest in developing policies that support ecologically sustainable and socially equitable cities. The 'technological fix', the idea that we can build, create, and innovate our way out of problems, eventually came to be seen as inadequate. The broader social, environmental and sometimes even economic ramifications of technological solutions were not fully considered in many of the project histories examined in the OMEGA project, part of the family of programmes to which the GAMUT project contributed.[2] The difficulty, however, was finding a way to bring these considerations to bear on policy making. A broader regard for the intended and unintended consequences of policy decisions resulted in the advance of social impact and ecological impact assessments in problem solving. While the process of impact analysis is quite prescriptive and mandates the engagement of the community, the results have often been paper-pushing exercises that give tokenistic regard to the impacts of development projects on society and the environment, and with little genuine engagement of the community.

This was often achieved by artful drafting of the briefs to exclude all but a limited set of the relevant options to a narrow range of alternatives, ensuring the preferred ones were appropriately managed through the final stages of the political process – restricting analysis and thus the scope of any later 'consultation'. The major East–West Link Study in Melbourne, Victoria (Eddington, 2008) arguably followed such procedures.

However, in pluralist western democracies, grassroots and community-based organizations have the right to mobilize outside formal bureaucratic structures in response to unpopular policy proposals. In the Eddington Report, an analysis was tabled signalling the need for a rail tunnel – and road tunnel – through the inner suburban areas of Melbourne but ignoring the need for rail extensions to outer growing suburbs.[3] Responding to the proposal, community groups including the Public Transport Users Association, the Melbourne Transport Forum and the Greens Party each delivered their own version of a transport plan that responded to the perceived inadequacies of the government proposal. Dissent towards the formal transport planning process also conjured responses from academic planners advocating particular transport policy responses that would also achieve broader social and environmental planning objectives (Low and Russell, 2008; Wigan, 2008. Documents 11.1 and 11.2 in the Companion Website).

The idea of epistemic communities has emerged to describe how groups of experts acquire power from elected representatives in a community context (as mentioned in Chapter 4, note 1). The concept has become useful in a number of areas, initially in international settings, but also describes how community and planning power structures may be better understood. A concise description of this useful concept is given by Haas (1992) and Sundstrom (2000). One challenge that deliberative democracy confronts is how to bring epistemic communities – including community interest groups – together to have the hard policy discussions necessary to address the complex challenges facing cities. Deliberative democracy theory posits that when policy making is conducted in an open, inclusive manner whereby power differences between participants are minimized by the nature of the process, the result is a policy outcome that is more legitimate and rational than the policy outcome that would have emerged from a more traditional and top-down process. Here the rationality of the outcome stems from face-to-face discussion among stakeholders about the problem being addressed. Deliberative democracy has a transformational quality that comes from the opportunity for participants to learn about the interests of others and to build trust. The key here is that conflicts of interest and perspective, such as arise from the dilemmas discussed in the first part of this volume, are not papered over with anodyne policy prescriptions. Instead they are fully confronted and worked through. The eminent political scientist Charles Lindblom, who we have cited on several occasions in this book, in his later work (1990) argued that all social inquiry suffers from 'impairment' – that is crucial omission of significant perspectives in the very attempt to focus on a problem. He posits, 'Possibly the most effective single way to reduce impairment is to get into circulation a greater variety of messages. Hence I turn again to the competition of ideas' (ibid.: 293). The process of reaching agreement itself becomes a social learning opportunity, and it is for this reason that social and political theorists advocate deliberative democracy for producing outcomes that go beyond individual interests to better reflect the public interest. Likewise, ecological theorists have posited that the outcomes of deliberative processes may produce a more ecologically rational outcome. It would seem then that open, deliberative processes are more likely to achieve sustainable transport outcomes than the traditional top-down rational planning.

The role of knowledge

One of the main tenets of deliberative democracy is that different forms of knowledge exist and each form of knowledge is bounded by experiences, technical knowledge and ideologies. Deliberative democracy is underpinned by a social constructivist epistemology in which knowledge is formed by experience and can be created, altered or expanded in interaction with the knowledge of others (Archer, 1998; Kukla, 2000; Flyvbjerg, 2001). It is in this process that meaning is given to knowledge (Healey, 2007: 238). An individual's knowledge is limited by many factors – for example, the breadth of their personal experience, expertise and social interaction. Therefore, without expanding this knowledge through face-to-face deliberation, the decisions actors make are constrained to a comparatively narrower scope (Healey, 2007).

When people are reluctant to change their position and insist on imposing their values by adopting stronger bargaining tactics, their position will be overshadowed by their rational individual interests at the expense of the collective good (Rawls, 2001: 191). For instance, challenges associated with instrumental rationality exist when actors make decisions based upon a view of cost-efficiency but lack consideration for the value, meaning and motivations behind what is being considered (Low, 2005: 48; Verma and Shin, 2005: 13). It is not uncommon for cost-efficiency analyses to neglect, understate or exclude factors regarded as important by the affected parties

A collectively rational dialogue occurs when the viewpoints of participants are subject to change as a result of the validity and strength of the better argument (Muller, 2004: 409; Risse, 2000: 12). Thus, rationality is achieved when reasons converge on a policy decision (Dewey, 1930: 196). From different theoretical traditions both Dewey and Habermas (1990) view rationality as stemming from free and open deliberation in a process that enables people to reflect on their own positions and values. These authors contend that such a condition is necessary to resolve complex social problems (Rosenthal, 2002: 210). In this respect Dewey offers what has been called an 'epistemological defence of democracy', arguing that without the participation of the public through consultation and discussion, political decisions cannot reflect the broader interests of society (Putnam, 2004: 104).

More radically, some proponents of deliberative democracy advocate removing influence from the traditional places where power is exercised and giving it to the people (Stevenson, 1988: 136). Others, however, posit that the public does not possess enough knowledge to make decisions advantageous to the future of the urban environment – the lay public has a 'cognitive deficit' (Eden, 1998; Wynne, 1991: 113). Another line of argument against transferring power to the wider public posits that politicians are elected to perform the role of making informed decisions, and they should therefore retain this policy-making responsibility. Why then, in a representative democracy, should citizens be continuously involved in policy making? (Fagence, 1977: 51). Furthermore, participation of the lay citizen, it can be argued, detracts from expert-based knowledge – or 'professional social inquiry' – acquired by professionals within their respective fields (Lindblom and Cohen, 1979: 13–4). However, as these authors recognize, when more than a quarter of the population possess tertiary qualifications a world has passed away where the 'expert knowledge' of the professionals in their bureaucratic silos is the only knowledge that counts. Greater levels of accountability are now demanded, but governments and their political leaders are understandably slow to respond to this sea change for reasons of path dependence and governmentality, as discussed in Chapter 6.

Embracing deliberation to minimize power play

The application of deliberative democracy in practice has provoked sceptics to argue that deliberative democracy is flawed because some stakeholders simply have louder voices (Sanders, 1997: 11) or are better positioned financially to have their voices heard (March and Olsen, 1989: 144). While some critics believe that power differentials between groups and individuals can destroy a process of argumentation, Habermas believes that pre-existing power differentials can be mitigated by setting rules for deliberation that remove or severely modify the potential for coercion or manipulation (Torgerson, 1999: 114–5). Such rules, Habermas claims, will enable a focus on the 'power of rationality' as the basis of decision making. Communication is, therefore, to be free of domination. It is not the validity of the claim that motivates an actor to accept an idea, but the offer to adduce reasons by the speaker that the hearer subsequently accepts (Habermas, 1990: 58–9).

Nevertheless pre-existing power structures pose a challenge for deliberative democracy. Flyvbjerg (1998: 194) argues that if power relations are not controlled, deliberation becomes a forum for antagonistic confrontation. Hence, if power differentials among interest groups and stakeholders are not addressed and neutralized, then democratic deliberations can be neither *deliberative* nor *democratic*. Much depends on the rules of engagement at any deliberative occasion. The rules should therefore emphasize inclusive and unconstrained dialogue to ensure that participants are not affected by domination and strategic interaction (Dryzek, 1990: 36). The 'ideal speech situation' advocated by Habermas can only occur when power is equalised among stakeholders. It requires the presence of a skilled facilitator with the authority to neutralize power differentials (Muller, 2004: 410 cites Haacke, 1996 and Holzinger, 2001). Moreover, unless information is also shared among the parties, such equalization is unlikely to occur. This is certainly the case in planning and transport matters where the power of information has become progressively more significant.

The question of power is further explored by Foucault who, like Habermas, accepts the view that the self is a creation of society, but argues, however, that society is not better off under these circumstances because constraints are imposed on members of society through patterns of acculturation (Rorty, 1989: 63). This argument presented by Foucault suggests that individuals, while restricted by the power of others, are also restricted by the social constraints they themselves internalize as they become members of the society. Rorty (1989: 63) argues that the concerns raised by Foucault can be mitigated through the emergence of liberal society equipped with institutions that can assist with society's collective self-improvement. These institutions, Rorty argues, provide opportunities for democratic deliberation institutionalized in a way that mitigates the coercive forces of more powerful individuals over others. Yet these institutions are often not adequate to counterbalance existing power structures. Thus we commonly see a reversal of the idea of 'evidence-based policy'. British experience of evidence-based policy making led to the term 'policy-based evidence', 'to describe the end result where government agencies filtered out information that was inconsistent with government policy' (Harding, 2008, cited in Wigan, 2012).

The postulated institutionalization of formal structures of deliberative democracy has been opposed by sceptics who suspect that such institutionalization is problematic. This apprehension exists due to the possibility that deliberative institutions could be 'co-opted' and absorbed by the state (Dryzek, 1987: 213). The possibility of co-option occurring is increased by the misuse of compromise – or premature compromise – when participating parties meet in the middle without allowing genuine problem solving to take place (Caspary, 2000: 38). Co-option is further exacerbated by entrenched power divisions that weaken the potential for effective conflict resolution (ibid.). In response to these concerns, Dryzek developed the theory of *discursive* democracy as an alternative way to conceptualize stakeholder engagement (Smith, 2003: 74–5). In discursive democracy the aim is to achieve representation of different discourses and views rather than representation of persons, interests or organizations (Dryzek, 2010: 43).

To sum up, then, a more nuanced view of deliberative democracy suggests that in practice it can only really be approximated. Nevertheless the underlying principles of inclusivity, power neutrality and rationality have real potential to add value to any democratic process. Deliberative democracy needs to be framed in the urban and transportation planning process not as a new form of democracy, requiring conventional forms of democracy to be disregarded, but as a fresh and essential component of democratic practice. It is a 'desirable ingredient' of democracy (Saward, 2000: 68) and provides a normative framework for openly considered decisions. Institutional structures that enable active participation by stakeholders do not therefore amount to a rejection of conventional forms of democracy, but rather, as Eckersley (2004: 2) argues, foreshadow a 'post-liberal' enhancement of democracy. A widely quoted example of this process is the use of deliberative democracy – in the guise of 'eDemocracy' – by the Scottish Parliament for ePetitioning (Macintosh, Adams, Whyte, and Johnston, 2007) and the development of youth policies (Macintosh, Robson, Smith, and Whyte, 2003), where the younger generations were found to be very comfortable with this line of development of political participation.

Deliberative democracy is also appropriate where objectivist definitions of 'rationality' may exclude groups in the community with strongly held values that are neither based on rational analysis nor backed by independently verifiable evidence. Arguably, many faith-based groups fall into this category. Yet the information imbalances emerging between government, commercial organizations and individuals are now posing a moral imperative for such faith-based organizations to take up a stronger and clearer role in decision making (Wigan, 2010, 2012). There is mounting evidence that community engagement at a substantial level can produce better-accepted – and often simply better – outcomes in strategic transport planning.

Synthesis of the dissemination and engagement paradigms

The theory reviewed above must be translated into workable processes, and, to do this, a substantial investment needs to be made in mutual understanding and respect of the very different ways in which the world is understood by those involved: outlooks shaped by the tasks different professional actors undertake as well as various political and community perspectives. Collisions between value systems and understandings are common in planning and transport, and if the magic word 'sustainable' is added, a whole new set of unfocused and mutually contradictory perspectives and unspoken assumptions and values are drawn in. Notwithstanding the principles posited in Chapter 4, the term sustainability is fertile ground for debates on a wide range of social values. It is possible to work towards a synthesis of these very different approaches, ironically by addressing higher levels of rationalist concepts and methods as a basic building block. Perhaps taking evidence-based policy seriously requires a healthy dose of rationalist engagement of a higher than usual quality?

The growing dependence on information technology by government is reflected in the initiation of serious research on how best to bring to bear on deliberative processes support for eGovernment. The lessons of successful eGovernment initiatives, and securing public engagement were learned early (Elmagarmid and McIver, 2001), but have not always been applied.

The EU Framework 7 IMPACT project[4] specifically addresses the issues of argumentation building and resolution, allied to textual analysis tools to make the potentially huge masses of text-based materials into a single usable deliberation framework. While this builds on successful eGovernance approaches (OECD Expert Group on Government Relations with Citizens and Civil Society, 2003), it is a move that adds further expert complexity – and by implication expert power – to the processes of policy formation and negotiation. The EU Framework 7 UBIpol project[5] focuses on delivering a suitable platform for very large numbers of participants, and the EU Framework 7 project CockPIT[6] addresses social and web 2.0 tools for engagement. Such advanced rationalist approaches will only be able to work if information access has been substantially improved and equalized by all the participating parties. Information sharing and transparency are the keys. The engagement process is the lock.

Process contexts

The deliberative and rationalist engagement models do not address another key aspect of information dissemination and process support: that of timing. There is a major temporal disjunction between the perspectives on consultation and engagement held, on the one hand, by the different professional disciplines involved in strategic – urban and transport – planning, and, on the other, by the communities involved and affected.

Strategic and urban planners, for the most part, consult carefully at the strategic planning stage, in a manner that is difficult for the affected communities to appreciate, understand and engage with. This is understandable, as such strategy determinants must be framed and settled a decade or more in advance of action on the ground. The specific implications of strategies for the communities affected cannot be grasped because the levels of detail are rarely in a form that can be readily engaged with by the communities, especially if the strategies do not relate to greenfield development. They are often, not surprisingly, expressed in visual and textual languages that the next generation of strategic planners are familiar and comfortable with.

Once the operational delivery occurs, the affected communities may have changed completely, and few of those present at the strategic consultation period would still remain. Direct building decisions involve disturbances and strong impacts on the ground during and after the construction phases, while the affected communities may have little or no memory of any consultation having taken place at all. Information alone cannot correct this deep divergence between the perceptions of engagement, but steps can be taken to soften the division. Visualization, involving computer-generated images of the end result of strategy is a basic tool, and often the endpoint of many aspects of urban and strategic development. Greater use of visualization holds some hope of bridging the community and professional divide between perspectives and stretching over time.

Excellent examples exist of visualization being brought to a greater and more comprehensive level of detail, accessible to a wider range of professional and community groups, in such landmark projects as the ETH/ Singapore Future Cities programme.[7] This programme also includes the detailed modelling processes used for the operational transport planning and other infrastructure requirements, thereby bringing the two very different times scales into the same context and timetable.

An example of dissemination response

The dissemination of knowledge in sustainable transport and planning requires both process and information components. The GAMUT programme recognized this, and at the point that a major project required considered input – the study of East–West transport infrastructure links across central Melbourne referred to earlier (Eddington, 2008) – the key feature was a discussion of the nature of consultation responses that GAMUT should submit, or be involved in (see Documents 11.1 and 11.2 in the Companion Website). The conclusion was that three responses were needed:

1 A conventional sustainability response supporting the elements that could be supported from that perspective.
2 An analytical review of the technical processes used and their reliability from the standpoint of the tools used and quoted in support of the study results (Wigan, 2008).
3 Catalysing the 'expert community' interests in an Alliance (Morris, 2008).

The first two were normal processes for a group with a diversity of skills and voices, but the third was a process not normally undertaken by a university centre. However, in terms of the dissemination of GAMUT findings it was a critical component. The transmission of process and of methods, under the auspices of a group such as GAMUT, was a major dissemination component in the drive towards sustainable transport. The community must take ownership of knowledge and information, and there is a special role for an academic group. In this role the group suspends advocacy of a particular stance. Instead the group takes on the task of enabling and supporting the base on which a wider community of stakeholders can combine their interests and secure evidence-based support for process, information and communication.

The framework for such a dissemination process took the form of a Governance Forum, where the GAMUT Centre at Melbourne University became host to diverse voices, without necessarily endorsing any or all of them, but providing a neutral territory on which such difficult and sensitive issues could be debated.

For GAMUT to adopt a neutral role as host of a forum may seem incompatible with the strong position on sustainability posited in Chapter 4, and the subsequent discussion of institutional barriers to change. Yet it is this very appreciation of the seriousness of the dilemma (discussed in Chapters 2 and 3) that makes it possible to respect the positions of those whose positions vary along a wide spectrum. Moreover, the acknowledgement of the structural nature of barriers to which all are subject (Chapters 5 and 6) encourages a position that does not seek to apportion blame to groups or individuals but rather invites mutual exploration of ways to overcome them.

The first such forum was the most crucial, and was set up to focus not on any past projects or initiatives, but to concentrate attention on the three key issues – based on the experience of each of the speakers – that should be addressed in the future. This simple framing allowed the Secretary of the Victorian Department of Transport to share a platform with a senior journalist, a GAMUT specialist in contestable evidence-based policy, a legal specialist in public–private partnerships (PPPs) and a director of a PPP road company, as well as a past Deputy Prime Minister. The attendance was entirely open, and those who accepted the invitation included members of the general public, local government politicians, consultants, policy analysts and research specialists. The forum proved to be effective in all the directions for which it had been designed, and a series of these forums followed (Legacy, Sturup, Curtis, and Wigan, 2009; Legacy, Curtis and Sturup, 2012). After the initial funding cycle for GAMUT ended, support was offered from government to continue to operate such forums for the dissemination and communication benefits of such a form of exposure of key issues that this model provides.

Where to next?

The role of both enhanced information sharing and wider and earlier engagement of expert and other communities in policy making continues to strengthen in importance. These two factors in combination enable evidence-based policy to move from something resembling a governmental battering ram to an active and effective engagement process stretching over a longer period than encompassed by the final negotiations around implementation.

Active moves to make strategic and tactical planning more effective require both information and processes to be modified so that options, visualizations and evaluations can be adjusted to secure a convergent understanding of alternative outcomes – and of constraints. This comprises a framework for *contestable* evidence-based policy (Wigan, 2011). Such an approach also requires better management of data, documents, mapping and other mechanisms for active sharing and decision-support using the many opportunities now afforded by ICT.

The growth of research in support of policy, and the increasing level of data handling and analysis that comes with it, have led to a consulting industry where the substantial evidence base of data and analysis required present a significant barrier to entrants within and outside government. Barriers to access to data may be of many kinds, but the top two are:

- privacy and undertakings made to subjects when data was collected
- commercial dependencies on datasets or tools.

The former requires careful protocols that will allow unit record data to be accessed by analysis tools, but where the third party analyst only gets the – audited – outputs of his tools. This protocol is now a central issue, as privacy has at long last become a matter of wide public concern.

The commercial and other issues are not limited to planning and transport (Napoli and Seaton, 2006). The same long-standing issues are identified of the abuse of policy research to legitimize pre-determined policy outcomes, and the use of research as political ammunition in both policy development (Albaek, 1995) and the subsequent advocacy (Rogers, 1989).

Good principles do not deliver good practice without support. While advanced tools can be useful, the evidence is that simple ones that are both understood and desired work even better (Elmagarmid and McIver, 2001). Barriers to sharing decision support systems such as transport and land use models can be substantial. These barriers are often simply the non-transparency of proprietary modelling systems, unaffordable costs, and a lack of training in their use. There are increasing numbers of extremely capable land use, geospatial, and transport modelling systems declared 'Open Source' – open to expert examination and critique – supported and developed by a worldwide community. Examples include TLUMIP from the State of Oregon in the USA (now SWIM2, although it is still extremely complex and slow to run), UrbanSim (www.urbansim.org) and MatSim (www.matsim.org). There are active communities supporting and applying both these latter models which are 'open source' and thus affordable and transparent. The main lesson from the long and extremely ambitious TLUMIP/SWIM2 programme at Oregon Department of Transport is that the key aspect is not the technical models themselves – however effective and sophisticated – but the manner in which they are used (Kudsen, 2011), and the engagement of all parties in this process.

The key factors that would enable such accessible tools to contribute to evidence-based policy, and to the broader transmission of the processes for deliberative democracy, are now in train at GAMUT (Sunter and Wigan, 2011). Exploring the vicissitudes – and hopefully also victories – experienced in trialling their potential contributions to contestable evidence-based policy in a deliberative democracy framework forms part of the ongoing programme for this work.

Transport and planning is but one of the fields where the increased use – and abuse – of knowledge and evidence-based policy debate has come under question (Napoli and Seaton, 2006). Data and analysis tools are essential elements in the conversion of classical dissemination methods to broader operational engagement by a wider range of stakeholder: community, expert and all levels of government included.

Notes

1 www.abp.unimelb.edu.au/gamut/publications-media/
2 See: http://www.omegacentre.bartlett.ucl.ac.uk/
3 http://www.transport.vic.gov.au/__data/assets/pdf_file/0011/34211/Investing_in_Transport-Overview-Part1.pdf (accessed 02/04/2012)
4 www.policy-impact.eu
5 www.ubipol.eu/UbiPOLHome.html
6 http://www.cockpit-project.eu/
7 www.futurecities.ethz.ch/sec/

12 Conclusion: dimensions of change

Nicholas Low

Introduction

In this book we have tried to present future urban transport as a global, long term, multi-dimensional problem. From wherever we stand on this planet it is today necessary to maintain this global, long term perspective. Whatever is occurring, whatever policies are adopted at particular times in particular places within their locally specific institutions, it is necessary to ask how this 'here and now' affects planetary ecosystems and future populations on this earth. Transport, dependent as it is today on burning fossil fuel, cannot be exempt from this perspective. Taking on this vision leads naturally to an interest in change, both technological change and social change, because the new science of global ecology tells us that change must occur, and quickly, if a benign and civilized world is to continue in prospect. It is not enough to observe change. It is necessary now to instigate deliberate socio-technical change. Of course, this is a philosophical problem that goes far beyond transport, but again the transport sector cannot be exempt.

The GAMUT research project on which the book is based adopted this perspective, and the result is a patchwork effort at best, with many pieces of the patchwork missing. The purpose of this chapter is to step back and look at some patterns in the patchwork. First the chapter shows how the different conceptual and theoretical approaches are woven together and related in order to reflect on the dilemma of mobility. For instance, turning points, critical junctures, discontinuities, or points of transformation are found within different theories of social change and stability. Second, the chapter reviews the different levels and domains in which the mobility dilemma and transport problem has been understood and investigated in this book: at global and city levels, and within technical, institutional and cultural domains.

Third, different experiences are revealed of governance and management of transport in the East Asia-Pacific region and areas of potential common learning. Finally the chapter brings together the solutions proposed to the dilemma of mobility, suggesting lessons for future governance of transport in different contexts, and reflecting on future research questions.

Theoretical frameworks

As we said in Chapter 1, Flyvbjerg's questions for social science as practical reason (or *phronesis*) inform our approach. Where are we going? Is this desirable? What should be done? But there is a prior question that we must also continue to ask: 'What is?' (*episteme*). Post-structuralism may have thought it disposed of this question, but it did not. Nevertheless science seeks answers through the lenses of theories. If it is hubris for natural scientists to imagine that there can be a single theory of everything, it is the height of arrogance for social scientists to do so. Different theories can illuminate reality in different ways, and these ways of illuminating reality also drive action so perhaps the *episteme-phronesis* distinction is also in need of revision. Be that as it may, we have used different theoretical frames to address the socio-technical problem of transport.

The book started by recognizing a dilemma – Marxists would say contradiction – within the market system of exploiting nature and human labour that we call 'the economy' (or 'capitalism'): the opposition between the current reality of physical mobility and the urgent need for ecological conservation. This dilemma will stimulate change, and the world is nearing a turning point. The question is not whether global change will occur, but whether the change will be benign and controlled, or destructive of economies and societies. There is a way out of the dilemma. We can probably have most of the mobility we want within ecological limits. But the mobility side of the dilemma is very concretely grounded in a material and social reality – private, fossil-fuelled vehicles and all the social and institutional supports that go with their use – as we sought to expose in Chapter 2 and again in Chapter 4. Since we cannot alter the ecological limits, the only way out is to change that material and social reality. One mistake that is made – often by transport specialists – is to assume that only technical change is necessary. But when we consider the scale of the necessary change that emerges from Chapters 2 and 3 it will be obvious that the social side poses a far greater difficulty. Another mistake, made by economists this time, is to assume that for change to occur it is only necessary to alter the fiscal and regulatory incentives – taxes and subsidies, emission caps and permits. Yet unless alternative means of mobility and accessibility are at the same time rapidly generated, change will be both painful and slow.

It may be thought that Chapter 2 is simply empirical. But by introducing space into understanding how the economy works the chapter adopts the theoretical perspective of geography, an indispensable perspective when thinking about transport – traversing space. Equally indispensable is the engineering perspective brought to bear on the transport problem in Chapter 3. In both chapters, however, the viewpoints developed are quite outside the box of conventional transport economics and engineering. Each takes a global perspective from different angles: Chapter 2 from the perspective of physical mobility in the global economy, Chapter 3 from the perspective of the ecological consequences of transport technology.

Chapter 4 was strongly shaped by the dilemma exposed in the preceding chapters. It continues the theme of social processes producing transport outcomes. But there is also a critique of political language: the language of 'sustainability' that spans from greenwash to moral imperative. The claim is that if sustainability is to provide a compass for social action then a rigorous and simple definition is necessary, and one that points unambiguously to the future rather than describing the present. This claim rests on an understanding of the now very extensive literature of the science of climate change, and of the appalling ecological and human risks attending a rise in global temperature in excess of 2 degrees Celsius (see for instance Lynas, 2008; Ward, 2008). The principles of sustainability posited in the chapter were supported by the theories of ecological economics (Daly, 1996) and environmental justice (Low and Gleeson, 1998).

Addressing the barriers to sustainable transport, Chapter 5 continued the theme of global capitalism as it developed in the twentieth century, and its changing regulation by governments. The framework employed here brings Karl Marx's path breaking analysis of nineteenth century capitalism closer to that of Karl Polanyi in the twentieth (Polanyi, 1957). The latter acknowledged the pivotal role of the state and its agents in creating and extending both the market principle and the social protection necessary to maintain the consent of the general public. The theory posits that change to the mode of regulation will occur at 'critical junctures', periods in which social consent becomes disrupted, and even violently withdrawn. These periods may see either a modification of the social provisions – in this case of transport – or a shift to a new mode of regulation. The experience in Japan is that both have occurred, though at different governance scales.

However, the broad global shift of state posture from Fordism to neoliberalism is not enough to explain the degree of institutional variability found in national and local societies. So 'path dependence' from institutional economics, the 'mentalities' of governance, and the philosophy of risk and affordance were explored to provide ways of understanding policy stasis or government 'business as usual'. The idea of 'critical juncture' appears again in path dependence theory. Here it is seen as a period when the context of action – when doubts occur and new ideas begin to gain traction, driving political mobilization and response from the state – favours the dislocation of a policy path. The research on the mentality of project told us that the window of opportunity for significant change to an infrastructure project – including dropping or postponing the project – slammed shut once the decision to proceed was authoritatively made.

The chapters that followed were more strongly empirical in focus but introduced a variety of theoretical frameworks within which to set their analyses. They also turned from *episteme* – explanation of what is – to *phronesis* – knowledge informing action. Chapter 7 described the very different governmental arrangements for transport in four cities of the Asia-Pacific region, and the different kinds of issues generated. Chapter 8 introduced regime theory's perspective on urban governance as well as the more institutional dimensions of political science (Leo, 1998; Stone, 1989; March and Olsen, 1996; Sabatier, 1999). Here the intention was to generalize from local experiences of 'challenge and response' to public policy, analysing both resistance to change, and strategies to overcome resistance. Effective political action to institute 'third order change' (Hall, 1993) demands local mobilization, which leads on to state capture – to a greater or lesser degree. The experiences in Japan described in Chapter 5 were instances of local spontaneous revolt, but in only one of the instances of revolt was there success in building longer term and more substantial change in government practices. These chapters reinforced the idea of 'windows of opportunity' – or critical junctures – but showed how actors were able to exploit these opportunities to bring about change.

The concept of deliberative democracy (Chapters 8 and 11) is typically a *phronetic* framework, combining as it does a value system and analytical strategy. In this work and also in Chapters 9 and 10, the perspective turns from the community trying to make change happen, to the policy shapers within government seeking change – policy entrepreneurs and planners. Sometimes community knowledge is enlisted to assist change of policy path dependencies, sometimes change is sought in community attitudes and behaviour. In the case of Chapter 9 the intention is to shift the focus of technical analysis – supporting political decision making – from the projection of demand from past behaviour embodying the value of market driven choice, to the analysis of present and future needs, embodying the underlying values of fairness, transparency and efficiency. The analysis revealed the potentials inherent in the land use and transport system, and set out the competencies required to manage public transport more effectively. In Chapter 10 the focus is on the shaping of automobile-dependent behaviour in childhood – leading to health and cognitive problems later in life. The chapter examines how that tendency might be changed through policies promoting children's independent mobility. Again the framework mixes values – the right of children to move safely in local space – with analysis in order to inform action.

Returning to the theme of democracy in transport planning, Chapter 11 expands the idea of deliberation beyond situations in which the community or its representatives engage in face-to-face talk with government officials. Certainly such forums in which power relations are temporarily suspended – as advocated by Habermas and others – have an important place. Discussion turns to the question of how pre-existing power relations might be suspended. The chapter broadens the scope of analysis to include a much wider spectrum in which policy relevant knowledge is generated and disseminated. The concept of 'epistemic communities' (Haas, 1992) is used to describe the connection between social organization and knowledge generation. The chapter brings within the scope of deliberative democracy both twentieth century modelling made possible by computer processing, and the application of information technology to democracy itself (eDemocracy).

What is the point of theory? It is not, as many students and some practitioners may suspect, to help academics fill the pages of scholarly journals. The point of theory is to enable generalization from locally situated and singular experiences. Theory is simply a level of abstraction that enables bridges to be built between experiences. That is as true for the social sciences as it is for the natural sciences. Without theory – and indeed values – all we are left with is a collection of descriptions of singular events which is of no use whatsoever in shaping policy de novo.

What does this mean for sustainable transport?

Transport planners and land use planners are social actors. What they think and what they do profoundly affects society, the quality of people's lives and the planetary environment. If the dilemma of mobility is to be progressively resolved planners need a broad understanding not only of the society-environment relationship, but of how that relationship is changing, and how it might be changed for the better. Social theory needs to be coupled with social values and social action. Today, the world is drifting towards a chaotic and uncontrollable outcome of the climate crisis. Short term fixes of economic woes and national security threats dominate the headlines. There is little debate and deliberation about the long term future. The large majority of the world's population is going to live in cities. The question of sustainable transport is ultimately a question of sustainable cities, and making cities of the future enjoyable places to live and work in, not merely places to tolerate.

It is no longer acceptable to project trends from past behaviour and provide for the continuation of that behaviour. Making cities and their transport systems sustainable requires values-based, future-oriented planning. The citizens of cities need to be engaged in that planning, however difficult that may be to achieve when citizens are beset with day-to-day problems. They need to be inspired with visions of better futures – especially at a time of economic depression. It should not be necessary for conditions to become so intolerable that citizens' only recourse is to violent revolt. But reinstating values-based planning will mean renegotiating the relationship between the power of citizens, their governments, and markets.

Levels and domains

Analysis in this book began with an examination of global structures of mobility reaching down to the intensive mobilities of city life and back again to their global consequences. An issue we encountered is that any serious consideration of mobility and transport must abandon any simple assumption of 'sedentarism' in geography and politics, the idea that there is somehow a natural tendency that fixes people and production to places. Of course it is true that people are not floating around like motes in some sort of global Brownian dance. Activities are still 'located' – they have a relatively stable locus or place. But it is increasingly true that people and products move around networks: the 'spaces of flows' (see Castells, 1996, and the collection of essays in Albrechts and Mandelbaum, 2005). So it becomes more difficult to place analytical boundaries around matters connected with transport. *Urban* transport is not really a separate category, distinct from regional, interregional, national or international transport – especially when it comes to examination of freight logistics. People may consume locally, but production comes from a global network. Even if the majority of people still live near their work, rapid transport technologies also mean that work can be a long way from home. How long distant relationships can continue to be connected by transport depends on whether technology can bridge the ecological gulf opening up today.

Even the dictum that 'all politics is local'[1] is beginning to be seriously challenged by the use of social networking technology. The most frequent opening remark on mobile phones is 'where are you?' But the government of states remains organized in spatial divisions. Global governance is weak, and where it occurs at all, is subject to the consent of nation states. There is, of course, no direct representation of the people at global level. Whether in the absence of global government global environmental problems can be solved effectively remains to be seen. Some are understandably horrified by the prospect of a global state, others are unimpressed by the outcome so far of the negotiations among competing nation states (Low and Gleeson, 1998: 189).

Always keeping in mind the global dilemma of mobility, the analytical approach in the book ranged from the international regulation of production to locally situated experience. The situated experience in the Japanese example was of crowds of people kept waiting on a station platform, resulting in the general and spontaneous feeling that this repeated experience had become intolerable. Threats to local quality of place or local amenity such as a public transport service gave rise to political mobilization (Chapter 8). So, political mobilization seems to begin around locally situated issues. In Chapter 10, the focus was on the factors that inhibited or enabled the growing range of roaming independently in space open to children as they developed into adulthood.

The analysis moved between the political action of the community – including 'stakeholders' and professionals outside the state – and of the state. All too often in neo-liberal times it is forgotten that the state and the government supply the power of the people collectively represented. Neo-liberal doctrine portrays the state as an alien power, an ectopic growth outside society, while the collective good is represented by the outcome of market transactions (Low, 1991: 166). The state can indeed be an alien power. But in democracies there are checks to prevent the worst excesses and misuse of state power from occurring. Further enlargement of the idea of democracy is necessary, but without the power of the state the will of the people is powerless in the face of the power of corporate business.

Scale and sustainable transport

Cities and transport systems are experienced by citizens in the city block, on the street, on the road, on the train; in short, locally. They may be good or bad experiences. But these local experiences often have roots extending into city-wide, regional, national and even global logistical and governance systems. So improving the local experience of city life will almost certainly require action at these higher levels and larger scales. Similarly, looking at the problem of change of transport systems towards low or zero carbon emission outcomes requires change right through the scales. If a nation imposes a price on carbon emissions, without the capacity at sub-national levels to provide infrastructure, and to structure incentives so as to change user behaviour, the desired response to that national policy setting will be feeble and slow. Simply, if people are not provided with a safe and efficient option to walk, or cycle, or take the bus or train, they will go on using their cars. The same will occur if incentives at different levels and in different domains – for instance in local car parking or the way taxes are applied – favour car travel. So alternatives must be available and incentives considered and applied consistently at all levels.

Local experiences also have roots in widely held social customs and beliefs such as how risk is evaluated, how children should be protected, and how problems should be solved. This is why the articulation in the modes of governance of relationships between geographical scales – localities, towns, regions, nations – and between domains of action expressed in organizations and institutions – transport, land use planning, engineering and construction, economics and social welfare – becomes so crucial to the movement towards sustainable cities and transport systems.

Governance structures

The chapters in this book raise two important questions. Are some governance structures more effective for sustainable urban transport than others? Are some governance structures more open to change than others? Note that the term 'governance' is used rather than 'government', the purpose being to include private sector organizations and ways of controlling them. Thus we have to consider three things: vertical organization between levels of government of different geographical scale and scope; horizontal organization between departments with different professional foci at the same level; and the division of responsibilities between the public and private sectors.

From Chapters 7 and 8 we can conclude that, to be effective, the governance of urban transport for a city requires city-wide, or even city-regional scope. Transportation itself is a physical network which, whether well connected or poorly connected, always remains a network or system of lines and nodes along which vehicles move. Automobility, the dominant use of private vehicles for movement, has given some weight to the notion that transport is nothing more than an agglomeration of individual transactions in a market for mobility. But that concept represents less than half the picture of city transport. First, no large city has been able to do without public transport in some form to keep automobile traffic from falling into terminal congestion. The more densely populated the city, the more necessary is public transport capable of moving large numbers of people quickly. Public transport is a public good, and attempts to fragment its planning and delivery among competing firms have had poor results. The public transport system needs to be planned and managed as an integrated whole, and private sector – or community sector – service delivery can function well as long as it is subject to the central planning of routes and timetables. The network benefits – of effective use of different modes supporting one another, of ease of transfer between modes, of simple integrated timetables, of whole of network fare structures, of promotion and marketing of public transport – cannot be achieved without centralized planning for the geographical territory of the network.[2] Where efficient public transport is available, incentives to discourage the use of private cars are also recommended – in addition to the effect on urban travel behaviour of pricing of carbon emissions.

In this respect, the contrast between system management in Zürich – and Perth – and that of Melbourne is very striking. In the latter case the recent history of public transport is one of incremental retreat from the fragmented management instituted by the introduction of a franchise system lacking central planning of routes and timetables. We might recall the advice of Nielsen et al. (2005: 50):

> The major institutional challenge for creating success in urban public transport is to combine the benefits of market competition and tendering in relation to market orientation and efficient service production, with the benefits of service coordination, network integration and the potential of public transport's economies of scale.

We should also remember, however, that there is an unavoidable trade-off between geographical service coverage and economic efficiency. The example of Japan shows that increased concern for efficiency – following privatization – resulted in thinning out of the public transport service coverage over Japan's territory (Chapter 5). But Chapter 9 presented a way of striking a more equitable balance between efficiency and service delivery over less densely occupied territory – in the New Zealand case study. Plainly there may also be ecological as well as social and economic implications of the particular trade-off. For instance, if the public transport system is fossil fuelled, the Fordist principle of providing a dense network of railways over the whole of a territory, irrespective of the intensity of its use, results in ecologically unsound use of energy and emissions. But if the system is electrically powered, with the electricity generated by renewable means, the ecological problem is much reduced or eliminated.

The second gap in the market mentality of automobility is that while vehicle movements may be individually governed in something like a mobility market, the road infrastructure is not. Major road systems are as much public systems as railways. They are almost always centrally planned – usually more so than railways – if sometimes privately financed. The problem here, however, is that road planning is often not well integrated with railway planning and land use planning. While the integration of public transport modes is necessary to form a seamless public transport system for a city, the integration of road planning with public transport planning appears to be more difficult. For effective integration to occur both modes need to be brought under the common direction of an urban or land use planning agency. However, structural change is necessary but not sufficient to ensure integration. Integration requires a change in the way the people in agencies or departments – or even sections of agencies – think, and that depends on their professional education. All transport and infrastructure planners need a broad education in urban planning, and likewise all urban planners need a well developed understanding of transport systems, their impact on cities, and how they can be best planned and managed. Once there is integration of professional knowledge in the minds of the staff of the agencies, it matters less how the agencies themselves are structured because the people who work in them will themselves seek integration between their activities and objectives.

Finally we come to the question of whether some structures of governance are more open to change and innovation than others. The argument pursued in Chapters 8 and 11 foregrounded processes: processes in which an alternative transport paradigm gained ascendancy, and deliberative and discursive processes through which a wider range of perspectives could be brought to bear on the dilemma of mobility. A background theme also concerned structures. In Chapter 8, it was noted that the Greater Vancouver Regional District was formed as a result of the community backlash against road policies predicated on automobility. The GVRD became a location for innovation, provided a structure for cooperation among local councils and undertook effective engagement with the public. In Perth, the Western Australia Planning Commission provided an alternative forum for urban planning questions to be discussed beyond the hierarchical structure of the WA State bureaucracy. The *Verkehrsverbund* model for public transport planning is likewise a cooperative structure – *Verbund* translates as 'community' – that enables the concerns of the stakeholders to be heard. By contrast, the tightly controlled and centralized bureaucratic planning of transport in Melbourne made it difficult for a new and alternative paradigm of transport planning to take effect. Instead there were incremental responses to external protests without paradigm change.

What we might draw from the structural background to these process-centred discussions is that one aim of change agents should be to create more open structures for planning future urban transport outside the central politically directed control of the state. The process undertaken by such structures can then approximate the deliberative and discursive methods in which a larger range of stakeholder groups can be included, and a larger range of perspectives can be brought to bear on the mobility dilemma and its attendant transport problems.

Governance for sustainable transport

Economist Ross Garnaut, in his review of climate change policy in Australia, identified a number of 'externalities' warranting 'corrective action' among them 'the positive externalities associated with new transport infrastructure and services' and 'the coordination externalities of integrated service provision (benefits accruing from integration of infrastructure, ticketing, provision of information and timing of services)' (Garnaut, 2008, section 19.5). Economists' language, based as it is on the assumption that markets are the normally perfect way to manage collective action, can thus become rather tangled. Put more simply, markets fail to provide the benefits of thoughtful and integrated planning of transport infrastructure and services. Why then should not thoughtful and integrated planning become the norm, and markets be employed to correct government failure? In fact, over the recent experience – during the past 200 years – of managing an increasingly integrated market economy, it is obvious that there can be both market and government planning failure. Both markets and planning are instruments for implementing the public good, and the art of good governance appears to be using each instrument to complement the other.

Every city will vary in the detailed ways in which markets and government planning are deployed. But there are some principles.

- Major benefits accrue from planning public transport systems as networks of different modes – heavy and light rail, buses, cycling and walking pathways.
- Such planning should be conducted by a single superior agency with responsibility for the future development of the city as well as its transport system – public and private aspects.
- The power of such an agency needs to be restrained by public accountability through continual and real public consultation in which deliberative public processes can play an influential role.
- Transport systems and their infrastructure require long term planning – with at least a 50-year time horizon – because infrastructure has a long life and imposes patterns on city development over a long period.
- Organizations with stakeholder representation are more likely to give rise to integrated solutions.
- The private sector may be involved in providing and managing infrastructure, but whether or not there is private sector involvement should be determined by the value to the public of reducing public risk or capital cost, or increasing efficiency and innovation.
- Transport infrastructure is a quasi-public good in that there is no necessary public benefit in being able to exclude from use those who do not directly pay. There is no inherent economic rationality in privatization or corporatization of transport infrastructure or operations.

Solutions

Automobility has become a real part of the structure of economies and societies worldwide. But these structures are dynamic, variable and changeable. The dynamism was captured in Chapter 4 in the image of the cyclone of automobility, and in Chapter 5 in the shifting mode of regulation of capitalism from Fordism to neo-liberalism. The variability was noted in Chapters 6, 7 and 8 with the observation that automobility has been significantly modified in some cities (for example, Zürich, Freiburg, Vancouver in Chapter 8) and implemented in different ways in others (for example, Melbourne, Perth, Singapore, Shanghai in Chapter 7). Transportation is also changeable. Automobility itself had to struggle for acceptance, and the treatment of transportation in the mode of regulation responds to pressures from the 'grassroots'. The need for governments to court consent from the people gives power to the people to require change. The question then becomes whether the demand from the public only occurs once conditions become intolerable (as in the Japan case studies in Chapter 5), or whether instead the demand can be generated by activists and policy makers offering an alternative vision *before* conditions become intolerable (as in the case of Vancouver in Chapter 8).

The dilemma described in Chapters 2 and 3 strongly suggests that transportation will change at every level and geographical scale in the next 50 years in response to the end of the fossil fuel economy and the neo-liberal mode of regulation. In Chapter 5 it was observed that a return to Fordism with its profligate expenditures on public services and infrastructure now appears impossible as nations struggle to contain or reduce debt. But sustainable transport now offers not only ecologically sound but also financially parsimonious solutions to mobility (see Jochem and Madlener, 2003; McKinsey and Co., 2008; Woodcock et al., 2009). Public transport for city commuting is less expensive per capita than commuting by car. Even though the expense of automobility is hidden in the private budgets of individuals rather than appearing on the public accounts of governments, the cost to the end user is just as real. Already in some countries, as fuel prices rise, commuters are voting with their feet to return to public transport modes. Cycling and walking is not only healthy but also cheaper transport. Improving infrastructure for these modes is many times less expensive than providing for private transport at high speed. As we proposed in Chapter 4, the positive 'quality of place' becomes an aspiration capable of engaging the imagination of the public.

In this book we have touched only lightly on the modes of electrical energy production for transport. It seems that the world may be at a critical juncture in this respect too. The Fordist modes of centralized energy production and networks of distribution have proved immensely expensive ways of generating and delivering electrical power to households and firms. The climatic risk of fossil fuelled generators is beginning to be accounted for – with carbon emission pricing. The promise of cheap nuclear energy has not been delivered. Not only are nuclear plants expensive to build and replace – and the nuclear waste to store – but the cost of the risks incurred from nuclear energy are simply not entered into the ledger. Perhaps in terms of cold calculation the number of deaths and injuries has so far been small compared with those incurred from automobility (see Chapter 3), but the possibility of irradiation frightens people, enough it seems for the Government of Japan to contemplate evacuating Tokyo after the meltdown of the Fukushima reactors following the great earthquake and tsunami of March 2011. Risk is informed by fear, and governments have to deal with that fear more than with the actual body count.

Rifkin (2009: 523) observes that 'transport revolutions are always embedded in larger infrastructure revolutions'. Drawing on several high tech industry sources, Rifkin foreshadows a 'distributed energy' future in which 'smart grids' allow electrical energy generated in homes, factories and offices to be exchanged and distributed to users with much reduced need for the back-up of centralized plants. Renewable energy is available almost everywhere in some form, whether wind, solar, small hydropower or capturing heat from earth and water. Smart grids and metering technology will 'allow local producers to sell their energy more effectively back to the main power grid as well as accept energy from the grid, making the flow of energy bidirectional' (ibid.: 521). The reconfiguration of the power grid 'along the lines of the Internet is now being tested by power companies in Europe, the USA, Japan, and China' (ibid.). Rifkin's point is that the transformation of energy systems will be driven by the need to economize: to get a better outcome from a smaller investment. So, sustainable transport is likely to be compatible with a post-liberal mode of regulation. Governments feeling the crunch between economy and legitimacy will have no alternative but to turn to ecologically sustainable modes of development to resolve their dilemmas, among them the dilemma of mobility.

While there are barriers to change (some are discussed in Chapter 6), there are also powerful social and economic forces generating change. If the term 'forces' seems too Marxist and abstract, we should also remember that the 'force' comes from people and ideas. Can change, then, be deliberately driven and shaped? Chapters 8 to 11 addressed this question. In Chapter 8, the complex processes of policy making were explored in which action from the grassroots comes to gain sufficient traction to overcome the resistance offered by established institutions. There must be opportunities for contestation in the local political arena. Challengers to orthodoxy must be able to put together coalitions of varied interests to promote new perspectives and accounts of problems, and exploit networks of cooperation and influence. They must also recognize critical junctures in which opportunities occur to mediate the impacts of global and local economic structures (as in Japan in Chapter 5 and Vancouver and Perth in Chapter 8). Such windows of opportunity open when a 'public crisis' occurs to which political leaders have to respond. However, real change probably requires that new people be installed in important bureaucratic and political positions. Such findings are consistent with Torfing's discussion of the impact of exogenous shocks and internal dilemmas in destabilizing path dependencies (Torfing, 2001: 288). Curtis and Low (2012; Chapter 12), provide a list of 'action guidelines' for change agents aiming to shift path dependent institutions of transport planning and policy.

It was also noted that examples where the new paradigm of sustainable transport had been most clearly adopted demonstrated a strong integration between transport and urban development policies. Chapter 9 went on to identify some pathways of analysis for land use and transport planning which are not dependent on the simple projection into the future of past travel behaviours. In this respect it is worth returning to the analysis in Chapters 2 and 3 pointing to a more sustainable land use and transport pattern: the polycentric city. It is not so much the overall population density of the city that is most significant in reducing travel demand, but rather the degree of polycentricity. This is certainly a postulate worth testing further. It is also consistent with the idea that quality of place needs to be achieved in the lower density suburbs of cities.

In Chapter 10, the place of transport and mobility in childhood development was explored. The analysis showed why limiting children's independent mobility creates a problem, and how local governments, working in the framework of State policy, in the State of Victoria, Australia, have responded to international and national agendas promoting 'child-friendly cities'. Local governments, it was shown, have proved more innovative than State agencies. Further progress, it was argued, required a focus on children as *current* rather than future citizens with rights of their own beyond the simple right to be protected from harm.

We have noted above the place of agency structures in facilitating change. Chapter 11's focus is mainly on the processes by which 'enlarged thought' (Benhabib, 1988) can be operationalized to find solutions to the mobility dilemma. The aim should be to enable a variety of perspectives from outside the conventional paradigm to interact with professional policy communities. It is just as important that in this process the perspectives of the planners and transport professionals are not excluded or sidelined. The dissemination of information is never a one-way process, but must be a two-way street in which all the participants are open to persuasion. The information technologies of the twenty-first century, collectively described as eDemocracy, arguably offer a potential to permit this to occur in ways that have yet to be widely institutionalized. As the GAMUT programme found, the concept of a platform for this wider discussion removed from the daily politics of transport, was found to be useful in this respect. In effect the GAMUT forums provided a temporary structure for something approaching enlarged thought to occur around transport issues.

Shifting investment from automobility to sustainable transport

Sustainable transport requires a radical shift in investment towards making high quality places throughout cities with paths for walking and cycling to bring people from homes to the nearby activities they seek, and fully networked, comfortable, fast and efficient public transport systems. Outer areas of cities must be connected to the public transport network as much as inner and denser areas. That investment shift will mean also winding down the vast road programmes that are sustained only by the momentum of the path dependent cyclone of automobility discussed in Chapter 4. The private car on public roads is an inefficient and costly method of mass transportation in cities. The clear lesson from these road building programmes is that they do not solve but only exacerbate traffic congestion. There will be cars in sustainable cities, powered by renewable fuels, but they will become a minority mode for mass travel.

In the developing world there will sometimes be a case for building better roads in those areas which lack good quality roads, but the target should be to provide safe mobility to all, not maximum speed for the few. Limiting motoring speed to levels consistent with the safety of other road users need not reduce the average speeds of travel throughout the road system. Networking public transport will increase its use by increasing the number of destinations readily accessible as shown in Chapter 9.

Where to from here?

The task for academic centres interested in researching and promoting sustainable transport is broad and multi-disciplinary. While the idea of future 'urban' transport – as with the Future Urban Transport programme of the Volvo Research and Education Foundations – is a reasonable limitation of scope when the time and resources available for study are scarce, the geographical boundaries of the urban are in fact rather fuzzy. Transport is a network that extends worldwide. The urban part may be the densest but it is connected seamlessly to the rest. If we include freight and logistics, it was evident from Chapter 2 that we are dealing with a worldwide web. If air travel is included, the urban impact is very significant but hardly separable from the international airways system and its behaviour. The same is true of sea transport, which generates specific problems from the port to multiple urban destinations and across regions. Even land passenger travel transcends the urban when efficient high-speed rail is included. So, for the future, the scope of researching sustainable transport should include passenger and freight transport, land, air and sea transport, urban and rural transport, intercity and international transport. We need to know the impact of this web of movement on the global environment.

There is already a large amount of research activity under way around the world into transportation with reduced carbon emissions. But the separate research agendas need to be connected up, and connected with governments and policy makers. A future integrated research programme might therefore look like a matrix in which separate themes are enabled constantly to inform one another and enter a dialogue with governments within their territory. The crosscutting themes are familiar enough. Five such themes are suggested below, but this is hardly a comprehensive list and more will be added. What is important now is to explore the links among research agendas and how the information generated can be made to help policy makers.

Identifying technological pathways

Sustainable transport research should now aim not for low or reduced carbon transport and mobility, but zero carbon transport. The age of fossil fuel is over. We know that technologies exist that can deliver such carbon outcomes for transport. But there is no 'clean slate' on which to inscribe new transport systems for cities and regions. The aim must be to identify developmental paths for new technologies starting from where we are now. We need to answer the question: how can existing transport systems be directed over time along new technological paths? That question requires answers from institutional and political research.

A key question concerns the relationship between static energy generation and transport. If there is greater reliance on batteries – or hydrogen – to store energy, how will the energy be generated and how will it be stored in vehicles? There will be no single path. Different technological possibilities need to be assessed, including their carbon footprints, their economic costs and benefits, and the speed at which change is likely to occur. There is a need for modelling capable of measuring the emission outcomes of different infrastructure projects, transport modes and – sections of – transport systems, for instance using methods already being developed (see Hickman, Ashiru and Banister, 2010, 2011).

Adapting the structure of industry

Transport is a major industrial sector throughout the developed and fast developing world. In Australia in 2008, the 'transport and storage' industries occupied 4.7 per cent of both national GDP and total employment. Half a million people work in the sector. The total value of transport infrastructure engineering construction in 2007–08 was $52.8 billion (BITRE, 2010). The industries include construction, concrete and bitumen, logistics, haulage, and ports, as well as air, rail and sea transport. All of these industries are currently based on the assumptions of costless carbon emissions and cheap fossil fuel. There is therefore a major issue of industrial restructuring ahead. We need to explore the nature of industrial change that the emerging sustainable transport pathways require, and the costs and benefits of adapting industry to the new set of assumptions around the need to reduce carbon emissions to zero, conserve energy and adapt to rising fuel costs.

Changing perceptions and behaviour

Transport is inscribed in our ways of life. Transforming transport therefore carries with it the possibility of both positive and negative outcomes for lifestyles. How people perceive these outcomes influences the public choice of developmental pathway. We need to describe the social consequences of different technological paths, for instance in social connections, employment opportunities and in public health. New transport technologies can help answer the question of how much mobility is sustainable, but we need also to explore the potential consequences of a low mobility society. What health benefits and costs might flow from greater reliance on public and active transport? (see Woodcock et al., 2009.)

Changing the law and public institutions

As with current transport technologies, the laws and institutions governing transport operations and infrastructure are based on twentieth century assumptions about the future of transport and mobility based on uncosted emissions and cheap fuel. Many fiscal and bureaucratic institutions are still geared to the assumption of cheap fossil fuel. We need to make transparent the hidden subsidies, via fiscal regimes, that privilege one form of transport over another. These subsidies and taxes act as powerful incentives so we need a thorough review of all laws and regulations pertaining to transport to reveal their positive or negative effects on the potential for transformation to sustainable transport.

Planning transport operations and infrastructure and shaping our cities

New technologies, if they are to succeed, require new infrastructure whose provision crosses governmental boundaries, both vertical – between levels of government – and horizontal – between government agencies. For instance electric cars will require recharge points whose provision is subject to local government planning regulation. Transfers between transport modes require well-designed intermodal nodes, both for freight and for passenger traffic. A priority for active transport requires infrastructure geared to safe cycling and walking in a pleasant environment. We need more examination of 'transit oriented cities', their design and implementation (see Curtis, Renne and Bertolini, 2009) and the possibilities inherent in existing patterns of urban development for encouraging a more polycentric form of development. We need better management of public transport systems (Mees, 2010).

Linkages

It is the interdisciplinary and inter-programme questions that are now most important to address. For instance:

- How do laws and institutions influence the use and development of modes and systems of transport, and how do they need to change to facilitate sustainable transport?
- What opportunities do zero carbon transport technologies open up for human aspirations and public health, and how can the benefits of new technologies be more fairly distributed?
- How does the structure of industry, and its ownership, influence transport technologies in use, and how does that need to change in order to respond to the need for sustainable transport?
- How do existing institutions shape the patterns of urban development, and how do patterns of urban development constrain or facilitate the opportunities afforded by new technologies?

There are doubtless many more such questions that will emerge when research programmes talk to one another, and talk with communities and governments as discussed in Chapter 11.

Conclusion

In the next 50 years transport will be transformed. But will it happen catastrophically as a crisis response, or benignly in a controlled and planned way? If the sort of questions we have posed seem hard to answer, and the solutions difficult to implement, the scope of the mobility dilemma is so great that no question and no solution can be thrown into the 'too hard' basket.

If, as with other matters of sustainability, we view transport and mobility as a zero sum game, the nations will continue to debate who was responsible for the problem, who must act first, who will lose most, who should pay, how action can rationally be postponed. This kind of debate is a smokescreen to disguise the magnitude of the problem and the failure to act decisively and collectively to deal with it. If, however, sustainable transport beyond the carbon era is viewed as serving the mobility needs of the future, then perhaps nations can begin to see the opportunities that open up for their peoples. Seen in this way all nations can and should do what some are already doing: discover their own paths to reach that future as quickly as possible, and before the world is locked irreversibly into a cycle of rapid global warming from which no nation will emerge a winner.

Notes

1 As a Speaker of the US House of Representatives is reputed to have once said.
2 If transport and logistical networks are theoretically boundless, it is still practical to draw boundaries around the denser parts of the networks, typically city regions, for purposes of their planning and management, thus making a pragmatic distinction between intra and inter-regional transport.

Bibliography

ABS (Australian Bureau of Statistics) (2008) *Perth: A Social Atlas*. Catalogue number 2030.5. Canberra.

—— (2011) *Regional Population Growth, Australia, 2009–10 Population Estimates: [Totals] by Geographic Classification [ASGC 2010], 2001 to 2010*. Report no. 3218.0.

Adams, J. (1999) 'The social implications of hypermobility' in OECD, Project on Sustainable Transportation (EST): *The Economic and Social Implication of Sustainable Transportation, Proceedings of the Ottawa Workshop, 20–21 Oct. 1998*, Paris: OECD Publications.

Adams, J. (2000) 'Hypermobility', *Prospect*, (March issue) 50: 27–31.

Aglietta, M. (1979) *A Theory of Capitalist Regulation: The U.S. Experience*, London: New Left Books.

Aguiléra, A. (2009) 'Employment suburbanisation, reverse commuting and travel behaviour by residents of the central city in the Paris Metropolitan Area', *Transport Research Part A: Policy and Practice*, 43: 685–691.

Akerman, J. and Hojer, M. (2006) 'How much transport can the climate stand? Sweden on a sustainable path in 2050', *Energy Policy*, 34: 1944–1957.

Albaek, E. (1995) 'Between knowledge and power: utilization of social science in public policymaking', *Policy Sciences*, 28: 79–100.

Albrechts, L. (2006) 'Shifts in strategic spatial planning? some evidence from Europe and Australia', *Environment and Planning A*, 38: 1149–1170.

Albrechts, L. and Mandelbaum, S. (eds) (2005) *The Network Society, A new context for planning*, London and New York: Routledge.

Aleklett, K., Hook M., Jakobsson, K., Lardelli, M., Snowden, S. and Soderbergh, B. (2010) 'The peak of the oil age – analysing the world oil production reference scenario in World Energy Outlook, 2008', *Energy Policy*, 38: 1398–1414.

Allsop, R. (2007) *Victoria's Public Transport: Assessing the Results of Privatisation*, Melbourne: Institute of Public Affairs.

Alpkokin, P., Black, J., Kato, H. and Vichiensan, V. (2007) 'Poly-centric employment formation in mega-cities: analysis from APEC-TR collaborative research', *Journal of the Eastern Asia Society for Transportation Studies*, 7: 1446–1459.

Altschuler, A., Anderson, M., Jones, D., Roos, D. and Womack, J. (1984) *The Future of the Automobile*, London: George Allen and Unwin.

Ameratunga, S., Hijar, M. and Norton, R. (2006) 'Road-traffic injuries: confronting disparities to address a global-health problem', *The Lancet*, 367: 1533–1540.

Amin, A., (ed.) (1995) *Post-Fordism: A Reader* (Studies in Urban and Social Change), Oxford: Blackwell.

Anderson, J. (1975) *Public Policy-Making*, New York: Praeger.

Andersen, J.L.E., Landex, A. and Nielsen, O.A. (2006) 'Light rail project in Copenhagen – the Ring 2½ corridor', paper presented at the Annual Transport Conference, Aalborg University, Denmark. http://www.trafikdage.dk/td/papers/papers06/Trafikdage-2006-556.pdf. (accessed 2 April 2011).

Angel, B.W.F., van Damme, D.A., Ivanovskaia, A., Lenders, R.J.M. and Weldhuijzen, R.S. (2006) *Warehousing Space in Europe: Meeting tomorrow's demand: a pan-European warehousing trends study by Capgemini and ProLogis*. Utrecht. Prologis and Cap Gemini Consulting Services. http://www.capgemini.com/m/en/tl/tl_Warehousing_Space_in_Europe__Meeting_Tomorrow___s_Demand.pdf. (accessed 15 July 2009)

Archer, M. (ed.) (1998) *Critical Realism: Essential Readings*, London and New York: Routledge.

Arendt, H. (1978) *The Life of the Mind: Willing*, New York: Harcourt Brace Jovanovich.

Arthur, W.B. (1988) 'Urban systems and historical path dependence', in J.H. Ausubel and R. Herman (eds) *Cities and Their Vital Systems, Infrastructure, Past, Present and Future*, Washington DC: National Academy Press.

Arthur, W.B. (1994) *Increasing Returns and Path Dependence in the Economy*, Ann Arbor, Michigan: University of Michigan Press.

Association for the Study of Peak Oil and Gas (ASPO) (2009) *ASPO Newsletter 100*, www.peakoil.net. (accessed 12 January 2010).

Auditor General (Government of Victoria) (2005) *Franchising Melbourne's Train and Tram System*, Melbourne: Victorian Government Printer.

Australasian Society for the Study of Obesity (2004) *Obesity in Australian Children Fact Sheet*, Sydney: Australian Society for the Study of Obesity (The Obesity Society).

Baer, P., Harte, J., Haya, B., Herzog, A.V., Holdren, J., Hultman, N.E. et al. (2000) 'Equity and Greenhouse Gas Responsibility', *Science*, 289: 22.

Banister, D. (2002) *Transport Planning*, London and New York: Taylor and Francis.

—— (2005) *Unsustainable Transport, City transport in the new century*, London and New York: Routledge.

Barter, P.A. (2008) 'Singapore's urban transport: Sustainability by design or necessity?' in T-C. Wong, B. Yuen and C. Goldblum (eds), *Spatial Planning for a Sustainable Singapore*, Dordrecht: Springer (pp. 95–114).

Bartlett, S. (2005) 'Integrating children's rights into municipal action: a review of progress and lessons learned', *Children, Youth and Environments*, 15(2): 18–40.

Bartlett, S., Hart, R., Satterthwaite, D., de la Barra, X. and Missair, A. (1999) *Cities for Children: Children's rights, poverty and urban management*, London: Earthscan.

Baur, L. (2007) 'The Millenium Disease', in G. Barnett (ed.) *The Millenium Disease: responses to Australia's obesity epidemic*. Canberra: Australian Parliament House: pp. 4–7.

Behrens, R. and Muchaka, P. (2011) 'Child independent mobility in South Africa: the case of Cape Town and its hinterland', *Global Studies of Childhood*, 3:167–183.

Bendixson, T. (1974) *Instead of Cars*, London: Temple Smith.

Benhabib, S. (1988) 'Judgment and the Moral Foundations of Politics in Arendt's Thought', *Political Theory*, 16(1): 29–51.

—— (1992) *Situating the Self: Gender, community and postmodernism in contemporary ethics*, Cambridge UK: Polity Press.

Berelowitz, L. (2005) *Dream City: Vancouver and the global imagination*, Vancouver: Douglas and McIntyre.

Bertolini, L. (1999) 'Spatial development patterns and public transport: the application of an analytical model in the Netherlands', *Planning Practice and Research*, 14: 199–210.

Betsil, M. and Bulkeley, H. (2007) 'Looking back and thinking ahead: a decade of cities and climate change research', *Local Environment*, 12: 447–456.

Bhat, C., Handy S., Kockelman, K., Mahmassani, H., Chen, Q. and Weston, L. (2000) *Development of an urban accessibility index: literature review*. Research project conducted for the Texas Department of Transportation. Center for Transportation Research, University of Texas, Austin (TX), USA.

BITRE (Bureau of Infrastructure, Transport and Regional Economics, Australian Government Department of Infrastructure and Transport) (2010) *Australian Transport Statistics Yearbook, 2009*, Canberra: Department of Infrastructure and Transport, www.bitre.gov.au/publications/2009/stats_001.aspx (accessed 10 December 2011).

Black, W.R. (2010) *Sustainable Transportation, Problems and solutions*, New York and London: The Guilford Press.

Boarnet, M., Anderson, C., Day, K., McMillan, T. and Alfonzo, M. (2005) 'Evaluation of the California Safe Routes to School legislation: urban form changes and children's active transportation to school', *American Journal of Preventive Medicine*, 28 (2 supplement): 134–140.

Böge, S. (1995) 'The well travelled yogurt pot: lessons for new freight transport policies and regional production', *World Transport Policy and Practice*, 1: 7–11.

Bontje, M. and Burdack, J. (2005) 'Edge cities, European-style: examples from Paris and the Randstad', *Cities*, 22: 317–330.

Bossel, U. (2006) 'Does a hydrogen economy make sense?' Proceedings IEEE 94, (10): 1826–1837.

Boyer, P. and Saillard, Y. (eds) (2002) *Regulation Theory: The state of the art*, London and New York: Routledge.

BP (2011) *BP Statistical Review of World Energy 2011*, London: British Petroleum.

Bratzel, S. (1999) 'Conditions of success in sustainable urban transport policy: policy change in "relatively successful" European cities', *Transport Reviews*, 19: 177–190.

Braun, B. and Castree, N. (1998) *Remaking Reality: Nature at the millennium*, London and New York: Routledge.

Breheny, M. (1999) 'Introduction', in M. Breheny (ed.) *The People: Where will they work?*. Report of TCPA research into the changing geography of employment, Town and Country Planning Association, London, pp. 1–8.

Brenner, N. (2004) *New State Spaces: Urban governance and the rescaling of statehood*, Oxford: Blackwell.

Bretagnolle, A., Paulus, F. and Pumain, D. (2002) 'Time and space scales for measuring urban growth', *Cybergeo: Revue Européenne de Geographie*, 219: 1–2.

Brown, J. and Thompson, G. (2008) 'Service Orientation, Bus-Rail Service Integration and Transit Performance: An Examination of 45 U.S. Metropolitan Areas', *Transportation Research Record, Journal of the Transportation Research Board*, vol. 2042, pp. 82–89.

Brueckner, J.K. (2003) 'Airline traffic and urban economic development', *Urban Studies*, 40: 1455–1469.

Bruegmann, R. (2006) *Sprawl: A compact history*, Chicago: University of Chicago Press.

Brugmann, J. (2009) *Welcome to the Urban Revolution: How cities are changing the world*, St Lucia: University of Queensland Press.

Buchanan, J.M. and Tullock, G. (1965) *The Calculus of Consent*, Ann Arbor, Michigan: University of Michigan Press.

Buehler, R. and Pucher, J. (2011) 'Making public transport financially sustainable', *Transport Policy*, 18: 126–138.

Burke, M., Evans, R. and Hatfield, E. (2008) *A Sporting Chance: Accessibility of proposed AFL stadium locations on the Gold Coast*, Brisbane, Australia: Urban Research Program, Griffith University. http://www.griffith.edu.au/environment-planning/urban-research-program/publications (accessed 10 January 2010).

Burke, M., Li, T. and Dodson, J. (2010). 'Exploring the Impacts of Employment Decentralisation in Brisbane using the Modular Urban Land Use and Transport Tool (MULUTT)', Paper delivered at the *GAMUT International Conference*. 2–4 June, Melbourne.

Burke, M. and Woolcock, G. (2009) 'Getting to the game: travel to sports stadia in the era of transit-oriented development', *Sport in Society: Cultures, Commerce, Media Politics*, 12: 890–909.

Button, K.J. (1993) *Transport Economics*, second edn., Aldershot, UK: Edward Elgar.

Button, K. and Stough, R. (1999) 'The importance of air transport for economic development', *Journal of Air Transport Management*, 106: 239–249.

Buxton, M. and Scheurer, J. (2005) *The Expanding Urban Frontier: Urban form in Melbourne's growth corridors*, Melbourne: RMIT Publishing.

Bwire, H. (2011) 'Children's independent mobility and perceptions of outdoor environments in Dar Es Salaam City', *Global Studies of Childhood*, 1: 184–204.

Cadzow, J. (2004) 'The Bubble-wrap Generation', *The Age*, Melbourne, *Good Weekend*: 18–22.

Callahan, R. (2007) 'Governance: The collision of politics and cooperation', *Public Administration Review*, 67: 290–301.

Campbell, C. (2003) 'The Peak of Oil, An economic and political turning point for the world', in N. Low and B. Gleeson (eds) *Making Urban Transport Sustainable*, Basingstoke, UK: Palgrave-Macmillan, pp. 42–64.

Campbell, J.C., Basin, M.A., Baumgartner, F.R. and Halpern N.P. (1989) 'Afterword on policy communities: a framework for comparative research', *Governance: An International Journal of Policy and Administration*, 2: 86–94.

Carrillo, F.J. (ed.) (2006) *Knowledge Cities: Approaches, experiences and perspectives*, New York: Elsevier Butterworth.

Casello, J.M. (2007) 'Transit competitiveness in polycentric metropolitan regions', *Transportation Research Part A: Policy and Practice*, 41: 19–40.

Caspary, W.R. (2000) Dewey on Democracy, London: Cornell University Press.

Castells, M. (1996) *The Rise of the Network Society*, Oxford: Blackwell.

Catalyst Communications (2003) *Cross City Tunnel: Summary of Contracts*, Sydney: Catalyst Communications.

Catford, J.C. and Caterson, I.D. (2003) 'Snowballing obesity: Australians will get run over if they just sit there', *Medical Journal of Australia*, 179: 577–599.

de Cerreño, A.L.C., Shin, H-S., Strauss-Wieder, A. and Theofanis, S. (2008) 'Feasibility of freight villages in the NYMTC region', Center for Advanced Infrastructure and Transportation Freight and Maritime Program. Rutgers University and New York University Rudin Center for Transportation and Management.

Cervero, R. (1998) *The Transit Metropolis: A global inquiry*, Washington DC: Island Press.

—— (2005) 'Progress in coping with complex urban transport problems in the United States' in G. Jönson and E. Tengström (eds) *Urban Transport Development, A Complex Issues*, Berlin: Springer, pp. 118–133.

CFPTVC (Committee on the Future of Personal Transport Vehicles in China) (2003) *Personal Cars and China*, Washington, DC: National Academies Press.

Chapman, L. (2007) 'Transport and climate change: a review', *Journal of Transport Geography*, 15: 354–367.

Chawla, L. and Heft, H. (2002) 'Children's competence and the ecology of communities: a functional approach to the evaluation of participation', *Journal of Environmental Psychology*, 22: 201–216.

Choi, C.C. and Toh, R. (2010) 'Household interview surveys from 1997 to 2008 – A decade of changing travel behaviours', *Journeys*, May: 52–61.

City of Ballarat (2007) *City of Ballarat: Becoming a Child and Family Friendly City – Report on Findings*, http://www.childfriendlycities.org/pdf/report_ballarat.pdf (accessed 16 October 2011).

City of Greater Geelong and David Lock Associates (2009) *Clause 56 Walkability Tool Kit: making walking preferable, not just possible*, http://www.dlaaust.com/downloadfiles/Walkability%20Toolkit_email.pdf, (accessed (23 April 2012).

City of Melbourne (2009) *Future Melbourne*, http://www.futuremelbourne.com.au/wiki/pub/FMPlan/PrintPlan/COM_SERVICE_PROD-%234771301-v1-Future_Melbourne_Exec_Summary_sml_pdf.pdf (accessed 16 October 2011).

City of Port Phillip (2005) *Creating a Child Friendly Port Phillip: Implementation Plan, 2005–2009*, http://www.portphillip.vic.gov.au/default/Implementation_Plan.pdf (accessed 16 October 2011).

City of Vancouver (2006a) *The Eco-Density Charter: Draft – Creating an Eco-City*, City of Vancouver, Vancouver, Canada.

—— (2006b) *Transportation Plan Progress Report*, City of Vancouver, Vancouver, Canada.

Clark, D. (1996) *Urban World, Global City*, London and New York: Routledge.

Conley, J. (2009) 'Automobile advertisements: the magical and the mundane', in J. Conley and A.T. McLaren, *Car Troubles, Critical studies of automobility and auto-mobility*, Aldershot, UK: Ashgate, Ch 2, pp. 36–57.

Conley, J. and McLaren, A.T. (eds) (2009) *Car Troubles, Critical studies of automobility and auto-mobility*, Aldershot, UK: Ashgate.

Considine, M. (2005) *Making Public Policy: Institutions, actors, strategies*, Cambridge UK: Polity Press.

Cooke, P. and Lazzeretti, L. (eds) (2008) *Creative Cities, Cultural Clusters and Local Economic Development*, Dordrecht and New York: Springer.

Costall, A. (1995) 'Socializing Affordances', *Theory and Psychology*, 5: 467–481.

Cox, P., Betts, R.A., Jones, C.D., Spall, S.A. and Totterdell, I.J. (2000) 'Acceleration of global warming due to carbon cycle feedbacks in a coupled climate model', *Nature*, 408: 184–187.

Cresswell, T. (2006) *On The Move, Mobility in the modern Western world*, London and New York: Routledge.

Crutzen, P.J., Mosier, A.R., Smith, K.A. and Winiwarter, W. (2008) 'N_2O release from agro-biofuel production negates global warming reduction by replacing fossil fuels', *Atmospheric Chemistry and Physics*, 8: 389–395.

Currie, G. and Senbergs, Z. (2007) 'Exploring forced car ownership in metropolitan Melbourne', Paper delivered at the *30th Australasian Transport Research Forum*, 25–27 September, Melbourne.

Curtis, C. (2005) 'The windscreen world of land use transport integration: experiences from Perth, a dispersed city', *Town Planning Review*, 76: 423–453.

Curtis, C. (2006) 'Network City: retrofitting the Perth metropolitan region to facilitate sustainable travel', *Urban Policy and Research*, 24: 159–180.

Curtis, C. (2012) 'Transitioning to transit oriented development: The case of Perth, Western Australia' *Urban Policy and Research*, 30.

Curtis, C. and Armstrong, R. (2009) *The Capacity of State and Local Government to Deliver Sustainable and Integrated Transport – Perth: Policy Position*, Working Paper no. 1. Melbourne: GAMUT, the University of Melbourne and Perth: Curtin University.

Curtis, C. and James, B. (2004) 'An institutional model for land use transport integration', *Urban Policy and Research*, 22: 277–297.

Curtis, C. and Low, N.P. (2012) *Institutional Barriers to Sustainable Transport*, Aldershot, UK: Ashgate.

Curtis, C. and Mellor, R. (2011) 'Anticipating a new railway: The relevance of transit oriented development to businesses in a car oriented city', *Urban Policy and Research*, 29: 1–26.

Curtis C. and Scheurer, J. (2009) 'Network City Activity Centres. Developing an Analysis, Conception and Communication Tool for Integrated Land Use and Transport Planning in the Perth Metropolitan Area'. Department of Planning and Infrastructure (DPI) and Curtin University of Technology, Perth (WA).

—— (2010) 'Planning for sustainable accessibility: developing tools to aid discussion and decision-making', *Progress in Planning*, 74: 2.

Curtis, C., Armstrong, R. and Babb, C. (2010) 'The capacity of state and local government to deliver land use transport integration: an analysis of land use and transport policies in Perth and Melbourne', paper delivered at the 12th World Transport Research Conference, Lisbon, June 2010.

Curtis, C., Renne, J. and Bertolini, L. (eds) (2009) *Transit Oriented Development, Making it happen*, Aldershot, UK: Ashgate.

Daly, H.E. (1992) *Steady State Economics*, Washington, DC: Island Press.

—— (1996) *Beyond Growth, Economics of sustainable development*, Boston: Beacon Press.

Davies, A.K. (2009) *The Structure of Suburban Employment in Melbourne*, unpublished PhD thesis. The University of Melbourne, Vic 3010, Australia.

Davison, G. with Yelland, S. (2004) *Car Wars: How the car won our hearts and conquered our cities*, Crows Nest (Sydney): Allen and Unwin.

Davison, G. (2004) *The Rise and Fall of Marvellous Melbourne*, 2nd edn. Melbourne, Australia: Melbourne University Press.

DCCEE (Department of Climate Change and Energy Efficiency) (2011) *State and Territory Greenhouse Gas Inventories 2009*, Canberra: DCCEE.

Dean, M. (1999) *Governmentality: Power and rule in modern society*, London, Thousand Oaks, CA: SAGE.

Dematteis, G. and Governa, F. (1999) 'From urban field to continuous settlements networks, European examples', in A.G. Aguilar and I. Escamilla (eds) *Problems of Megacities: Social inequalities, environmental risk and urban governance.* Institute of Geography, Universidad Nacional Autonoma de Mexico, UNAM and Commission of Urban Development and Urban Life, IGU, pp. 543–556.

Dempsey, P.S., Goetz, R. and Szyliowicz, J.S. (1997) *Denver International Airport: Lessons learned*, New York: McGraw-Hill.

Deng, F. (2011) 'Make road safety the new traffic mantra', *China Daily*, November 3rd p. 9 (the author is a senior transport specialist at the World Bank, Washington, DC).

Department of Infrastructure (DoI) Victoria (2002) *Melbourne 2030: planning for sustainable growth*, Melbourne, Australia: Department of Infrastructure, Government of Victoria.

Derudder, B., Devriendt, L. and Witlox, F. (2010) 'A spatial analysis of multiple airport cities', *Journal of Transport Geography*, 18: 345–353.

Dewey, J. (1930) *Human Nature and Conduct*, New York: The Modern Library.

Dicken, P. (1992) *Global Shift: The internationalization of economic activity*, Thousand Oaks, CA. SAGE Publications Inc.

DiGuiseppi, C., Roberts, I. and Li, L. (1997) 'Influence of changing travel patterns on child injury death rates', *British Medical Journal*, 314: 710–713.

Dimitriou, H. (1992) *Urban Transport Planning, A developmental perspective*, London: Routledge.

—— (2011) *Urban Transport in the Developing World, A handbook of policy and practice*, Cheltenham, UK: Edward Elgar.

Dodson, J. and Sipe, N. (2007) 'Oil vulnerability in the Australian city: assessing socioeconomic risks from higher urban fuel prices', *Urban Studies*, 44: 37–62.

Dowding, K. (1995) 'Model or metaphor? A critical review of the policy network approach', *Political Studies*, 43: 136–158.

Downs, A. (2004) *Still Stuck in Traffic: coping with peak hour traffic congestion*, Washington, DC: Brookings Institution Press.

DPUD (Department of Planning and Urban Development, Western Australia) (1990) *Metroplan: A Planning Strategy for Perth Metropolitan Region*, Perth, WA: Department of Planning and Urban Development.

Driandra, R.P. and Kinoshita, I. (2011) 'Danger from traffic to fear of monkeys: children's independent mobility in four diverse sites in Japan', *Global Studies of Childhood*, 1: 224–240.

Dryzek, J. (1987) *Rational Ecology: Environment and political economy*, Oxford: Basil Blackwell Inc.

—— (1990) *Discursive Democracy: Politics, policy and political science*, Cambridge: Cambridge University Press.

—— (2010) *Foundations and Frontiers of Deliberative Governance*, Oxford and New York, Oxford University Press.

Dudley, G. and Richardson, J. (2000) *Why Does Policy Change? Lessons from British transport policy 1945–99*, London and New York: Routledge.

Eckersley, R. (2004) *The Green State: Rethinking sovereignty and democracy*, Cambridge: MIT Press.

ECMT (European Conference of Ministers of Transport) (2003) *Implementing Sustainable Urban Travel Policies*, Paris: European Conference of Ministers of Transport.

Eddington, R. (2008) *Investing in transport: Overview East West Link needs assessment*, Melbourne: Department of Infrastructure, Government of Victoria.

Eden, S. (1998) 'Environmental Issues: knowledge, uncertainty and the environment', *Progress in Human Geography*, 22: 425–432.

EIA (Energy Information Administration) (2011) *International Energy Outlook 2011*, Washington, DC: US Department of Energy.

Elmagarmid, A.K. and McIver, W.J.J. (2001) 'The ongoing march toward Digital Government', *Computer*, (Feb): 32–38.

Engwicht, D. (1992) *Towards an Eco-City: calming the traffic*, Sydney: Envirobook.

Fagence, M. (1977) *Citizen Participation in Planning*, Oxford, UK: Pergamon Press.

Felix, E. and Tilley, D.R. (2009) 'Integrated energy, environmental and financial analysis of ethanol production from cellulosic switchgrass', *Energy*, 34: 410–436.

Fenton, M. (2003) 'Promoting walking in the US: overcoming the "stickiness" problem', in R. Tolley (ed.) *Sustainable Transport, Planning for walking and cycling in urban environments*, Boston: CRC Press and Cambridge UK: Woodhead Publishing, 41: 550–563.

Fincher, R. and Iveson, K. (2008) *Planning and Diversity in the City*, New York: Palgrave MacMillan.

Flink, J. (1988) *The Automobile Age*, Cambridge, Mass: MIT Press.

Flöbel, F., Heinrichs, J., and Kreye, O. (1980) *The New International Division of Labour*, translated by P. Burgess, Cambridge UK: Cambridge University Press.

Florida, R. (2005) *Cities and the Creative Class*, New York and London: Routledge.

Florida, R., Gulden, T. and Mellander, C. (2008) 'The rise of the mega-region', *Cambridge Journal of Regions, Economy and Society*, 1: 459–476.

Flyvbjerg, B. (1998) *Rationality and Power: Democracy in practice*, Chicago: The University of Chicago Press.

—— (2001) *Making Social Science Matter, Why social inquiry fails and how it can succeed again*, Cambridge UK: Cambridge University Press.

Flyvbjerg, B., Bruzelius, N. and Rothengatter, W. (2003) *Megaprojects and Risk: An anatomy of ambition*, Cambridge UK: Cambridge University Press.

Foo, T.S. (2000) 'An advanced demand management instrument in urban transport: electronic road pricing in Singapore', *Cities*, 17: 31–49.

Foucault, M. (1991) 'Governmentality', in G. Burchell, C. Gordon and P. Miller (eds) *The Foucault Effect, Studies in Governmentality with Two Lectures by and an Interview with Michel Foucault*, Hertfordshire: Harvester Wheatsheaf, pp. 87–104.

Freestone, R. and Murphy, P. (1998) 'Metropolitan restructuring and suburban employment centers: cross-cultural perspectives on the Australian experience', *Journal of the American Planning Association*, 64: 286–297.

Friedman, M. and Friedman, R. (1980) *Free to Choose*, New York: Harcourt Brace Jovanovich.

Friedman, T. (2005) *The World Is Flat: A brief history of the twenty-first century*, NewYork: Farrar, Straus and Giroux.

Funabashi, H. (1988) 'Kozoteki Kincho no Rensateki Suii', in Funabashi et al., *Regional Problems of High-Speed Civilisation: Construction and Conflict of Tohoku Shinkansen and Their Social Effects*, Yuhikaku (in Japanese).

Fyhri, A., Hjorthol, R., Mackett, R., Fotel, T. and Kyttä, M. (2011) 'Children's active travel and independent mobility in four countries: development, social contributing trends and measures', *Transport Policy*, 18: 703–710.

Garcia-Palomares, J.C. (2010) 'Urban Sprawl and the travel to work: the case of metropolitan area of Madrid', *Journal of Transport Geography*, 18: 197–213.

Gardiner, S.M. (2010) 'Ethics and climate change: an introduction', *WIREs Climate Change*, 1: 54–66.

Garnaut, R. (2008) *The Garnaut Climate Change Review: Final Report*, Port Melbourne, Australia, 3207: Cambridge University Press.

Garrard, J. (2008) *Safe Speed: promoting walking and cycling by reducing traffic speed*, Melbourne: Safe Speed Interest Group/ Heart Foundation of Australia.

—— (2009) *Active Transport: children and young people, an overview of recent evidence*, Melbourne: Victorian Health Promotion Foundation.

Garrard, J., Crawford, S. and Godbold, T. (2009) *Evaluation of the Ride2School program: final report*, Melbourne: Deakin University/Bike Victoria.

Garreau, J. (1991) *Edge City: Life on the new frontier*, New York: Doubleday.

Garrison, W.L. and Levinson, D.M. (2006) *The Transportation Experience, Policy, planning and deployment*, Oxford and New York: Oxford University Press.

Gaschet, F. (2002) 'The new intra-urban dynamics: suburbanisation and functional specialisation in French cities', *Papers in Regional Science*, 81: 63–82.

Gaymer, S. and Kinnear, R. (2009) 'Understanding recent changes in public transport patronage', *Department of Transport, Victoria, Australia: Transport Research and Policy Analysis Bulletin*, 6: 7.

Gehl, J. and Gemzoe, L. (2003) 'Winning back public space' in R. Tolley (ed.) *Sustainable Transport, Planning for walking and cycling in urban environments*, Boston: CRC Press and Cambridge UK: Woodhead Publishing, Ch 8, pp. 97–106.

Georgescu-Roegen, N. (1971) *The Entropy Law and the Economic Process*, Cambridge, Massachusetts: Harvard University Press.

Geurs, K.T. and van Eck, J.R.R. (2003) 'Evaluation of accessibility impacts of land-use scenarios: the implications of job competition, land-use and infrastructure developments for the Netherlands', *Environment and Planning B: Planning and Design*, 30: 69–87.

Gibson, J.J. (1986) *The Ecological Approach to Visual Perception*, New York: Psychology Press.

Gill, T. (2005) *Cycling and Children and Young People*, London: National Children's Bureau.

—— (2007) *No Fear: Growing Up in a Risk-Averse Society*, London: Calouste Gulbenkian Foundation.

Ginnell, K., Smith, P. and Oberlander, H.P. (2008) 'Making biggest bigger: port metro Vancouver's 21st century re-structuring – global meets local at the Asia Pacific gateway', *Canadian Political Science Review*, 2: 76–92.

Giuliano, G. and O'Brien, T. (2009) 'Responding to Increasing Port-related Freight Volumes: Lessons from Los Angeles/Long Beach and other US Ports and Hinterlands, in Port Competition and Hinterland Connections'. OECD/ITF Round Table 143. Paris. OECD/ITF. Pages 77–109.

Giuliano, G. and Small, K.A. (1999) 'The Determinants of Growth in Employment Subcentres', *Journal of Transport Geography*, 7: 189–202.

Glaeser, E.L and Kahn, M.E. (2001) 'Decentralized employment and the transformation of the American city', National Bureau of Economic Research Working Paper No 8117.

Glover, L. (2010) *Signs of Success or Signals of Failure? Symbolic politics and Melbourne's public transport privatisation*, Melbourne: GAMUT Working Paper (Australasian Centre for the Governance and Management of Urban Transport), University of Melbourne.

—— (2011) 'Public Transport as a Common Pool Resource', *Proceedings of the Australian Transport Research Forum Conference*, 28–30 September 2011, Adelaide. http://www.patrec.org/atrf.aspx (accessed 11 August 2012).

de Goei, B., Martijn J., Burger, M.J., Van Oort, F.G. and Kitson, M. (2010) 'Functional polycentrism and urban network development in the greater south east, United Kingdom: evidence from commuting patterns, 1981–2001', *Regional Studies*, 44: 1149–1170.

Gogia, N. (2006) 'Unpacking corporeal mobilities: the global voyages of labour and leisure', *Environment and Planning A*, 38: 359–375.

Goodwin, P.B. (1996) 'Empirical evidence on induced traffic, a review and synthesis', *Transportation*, 23: 35–54.

Goodyear, S. (2011) 'When Design Kills: the criminalization of walking', *Grist*, 20 July 2011. Available http://www.grist.org/infrastructure/2011-07-20-when-design-kills-the-criminalization-of-walking (accessed 18 October 2011).

Government Media Office (2001) *Government overhauls Western Australia's planning framework*, Ministerial Media Statements, Perth, WA.

Government of British Columbia (2003) *The Gateway Program Home*, Vancouver: Government of British Columbia. http://www.gatewayprogram.bc.ca/. (accessed 15 October 2010).

Gramsci, A. (1971) *Selections from the Prison Notebooks*, Q. Hoare and G. Nowell-Smith (eds), New York: International Publishers.

Guth, D., Holz-Rau, C. and Maciolek, M. (2009) 'Employment suburbanisation and commuter traffic in German city regions', paper delivered at the Swiss Transport Research Conference 2009, http://programm.corp.at/cdrom2010/papers2010/CORP2010_19.pdf. (accessed 17 September 2010).

Gutiérrez, J. and Gómez, G. (1999) 'The impact of orbital motorways on intra-metropolitan accessibility: the case of Madrid's M–40', *Journal of Transport Geography*, 7: 1–15.

GVRD (Greater Vancouver Regional District) (1975) *The Livable Region 1976–86: Proposals to manage the growth of Greater Vancouver*, Vancouver, Canada: Greater Vancouver Regional District.

—— (1993) *A Long-Range Transportation Plan for Greater Vancouver*, Vancouver, Canada: Greater Vancouver Regional District.

—— (1996) *Livable Region Strategic Plan, Greater Vancouver Regional District*, Vancouver, Canada: Greater Vancouver Regional District.

GVTA (Greater Vancouver Transportation Authority) (2004) *Greater Vancouver Region Trip Diary Survey, 2004*, Greater Vancouver Transportation Authority (TransLink), Vancouver.

Haacke, J. (1996) 'Theory and Praxis in International Relations: Habermas, Self-Reflection, Rational Argumentation', *Millennium Journal of International Studies*, 25: 255–286.

Haas, P.M. (1989) 'Why collaborate? Issue linkage and international regimes', *World Politics*, 32: 357–405.

—— (1992) 'Introduction: epistemic communities and international policy coordination', *International Organization*, 46: 323–365.

Habermas, J. (1990) *Moral Consciousness and Communicative Action*, Cambridge, Massachusetts: The MIT Press.

Hajer, M. (1995) *The Politics of Environmental Discourse, Ecological modernisation and the policy process*, Oxford: Clarendon Press.

Hall, P. (1993) 'Policy paradigms, social learning and the state: the case of economic policy-making in Britain', *Comparative Politics*, 25: 275–294.

Hall, P.V. and Clark, A. (2010) 'Maritime ports and the politics of reconnection', in D. Desfor, J. Laidley, Q. Stevens and D. Schubert (eds) *Transforming Urban Waterfronts: Fixity and flow*, London and New York: Routledge, pp. 17–34.

Hall, P.G. and Pain, K. (2006) *The Polycentric Metropolis: Learning from Mega-city Regions in Europe*, London and Sterling, VA: Earthscan.

Hamilton, K. (2003) 'If public transport is the answer, what is the question?' in A. Root (ed.) *Delivering Sustainable Transport, A social science perspective*, Amsterdam: Pergamon (Elsevier Science).

Han, S.S. (2000) 'Shanghai between state and market in urban transformation', *Urban Studies*, 37: 2091–2112.

—— (2005) 'Global city making in Singapore: A real estate perspective', *Progress in Planning*, 64(2): 64–175.

—— (2009) 'Managing motorization in sustainable transport planning: the Singapore experience', *Journal of Transport Geography*, 18: 314–321.

Hansen, H.H. and Winther, L. (2007) 'The spaces of urban economic geographies: industrial transformation in the outer city of Copenhagen', *Geografisk Tidsskrift, Danish Journal of Geography*, 107: 45–54.

Hansen, J.E., Kharecha, P., Sato, M., Ackerman, F., Hearty, P.J. and Hoegh-Guldburg, O. (2011) 'The case for young people and nature: a path to a healthy, natural, prosperous future', available at http://arxiv.org/ftp/arxiv/papers/1110/1110.1365.pdf. (accessed 17 October 2011).

Hao, H., Wang, H. and Ouyang, M. (2011) 'Comparison of policies on vehicle ownership and use between Beijing and Shanghai and their impacts on fuel consumption by passenger vehicles', *Energy Policy*, 39: 1016–1021.

Harcourt, M. and Cameron, K. (2007) *City Making in Paradise: Nine decisions that saved Vancouver*, Vancouver: Douglas and MacIntyre.

Harding, D. (2008) 'Fuel Watch evidence runs on empty', *The Age*, Melbourne, Australia 2 July.

Hart, R. (2002) 'Containing children: some lessons of planning for play from New York City', *Environment and Urbanization*, 14: 135–148.

Harvey, D. (1982) *The Limits to Capital*, Chicago: University of Chicago Press.

—— (1985) *The Urbanization of Capital*, Oxford: Blackwell.

Hasegawa, K., Funabashi, H. and Hatanaka, S. (1988) 'Tohoku-Joetsu Shinkansen no Kensetsu to Chiiki Funso', in H. Funabashi (ed.) *Regional Problems of High-Speed Civilisation: Construction and Conflict of Tohoku Shinkansen and Their Social Effects*, Yuhikaku (in Japanese).

Hauswirth, D.B., Hoffman, D.W., Kane, J.W., Ozobu, I.L., Thomas, C.L. and Wong, P.W. (eds) (2004) 'Alameda Corridor', in *Collaborative Leadership: Success Stories in Transportation Mega Projects*, U.S. Department of Transport, Federal Highway Administration, pp. 37–43. http://www.fhwa.dot.gov/ipd/pdfs/collaborative.pdf (accessed 15 August 2009).

Hayuth, Y. (1981) 'Containerization and the load center concept', *Economic Geography*, 57: 160–176.

Healey, P. (1997) 'The Revival of Strategic Spatial Planning in Europe', in P. Healey, A. Khakee, A. Motte and B. Needham (eds) *Making Strategic Spatial Plans: Innovation in Europe*, London: UCL Press Limited, pp. 3–20.

—— (2007) *Urban Complexities and Spatial Strategies: Towards a relational planning for our times*, London and New York: Routledge.

Heft, H. (2001) *Ecological psychology in context: James Gibson, Roger Barker and the legacy of William James's radical empiricism*, Mahwah, NJ: L. Erlbaum.

Heinz, J.P., Laumann, E.O., Nelson, R.L. and Salisbury, R.H. (1993) *The Hollow Core: Private interests in national policy making*, Cambridge MA: Harvard University Press.

Held, D. (1987) *Models of Democracy*, Cambridge UK: Polity Press.

Henderson, M. and Hogarth, F. (2007) 'Evaluation framework for assessing E-Democracy policy', in A. Anttiroiko and M. Malkia (eds) *Encyclopedia of digital government*, Hershey, PA: IGI Global, pp. 790–797.

Hesse, M. (2004) 'Logistics and freight transport policy in urban areas: a case study of Berlin-Brandenburg, Germany', Research Briefing European Planning Studies, 12: 1035–1053.

Hickman, R., Ashiru, O. and Banister, D. (2010) 'Transport and climate change: simulating the options for carbon reduction in London, *Transport Policy*, 17: 110–125.

—— (2011) 'Transitions to low carbon transport futures: strategic conversations from London and Delhi', *Journal of Transport Geography*, 19: 1553–1562.

Hidde, P., van der Ploeg, H.P., Merom, D., Corpuz, G. and Bauman, A.E. (2008) 'Trends in Australian children traveling to school 1971–2003: burning petrol or carbohydrates?', *Preventive Medicine*, 46: 60–62.

Hillman, M. (2002) 'Are we developing battery-reared or free-range children?' Presentation delivered at the Western Australia Pedestrian Advisory Council Seminar, Perth, March 14, 2002.

Hillman, M. (2003) 'The relevance of climate change to future policy on walking and cycling', in R. Tolley (ed.) *Sustainable Transport, Planning for walking and cycling in urban environments*, Boston: CRC Press and Cambridge UK: Woodhead Publishing, Ch 2, pp. 20–31.

Hillman, M., Adams, J. and Whitelegg, J. (1990) *One False Move, A study of children's independent mobility*, London: Policy Studies Institute.

Holzapfel, H. (1995) 'Potential forms of regional economic co-operation to reduce goods transport', in I.J. Koppen and H.A. Arp (eds) *World Transport Policy and Practice*, 1–2: 34–39.

Horton, D., Rosen, P. and Cox, P. (eds) (2007) *Cycling and Society*, Aldershot, UK: Ashgate.

Howe, A., Glass, G. and Curtis, C. (2009) 'Retrofitting TOD and managing the impacts: The case of Subi Centro', in C. Curtis, J.L. Renne and L. Bertolini (eds) *Transit Oriented Development: Making it Happen*, Aldershot, UK: Ashgate, pp. 65–73.

Humphreys, M. (2011) *Sustainability in European Transport Policy*, London and New York: Routledge.

Imran, M. (2010) *Institutional Barriers to Sustainable Transport in Pakistan*, Oxford UK: Oxford University Press.

Infrastructure Partnerships Australia (2006). *Case Studies – City Link Melbourne*. Melbourne.

Interrante, J. (1983) 'The Road to Autopia' in D.L. Lewis and L. Goldstein (eds) *The Automobile and American Culture*, Ann Arbor: University of Michigan Press.

IEA (International Energy Agency) (2010a) Key World Energy Statistics 2010, Paris: IEA/OECD,

—— (2010b) Oil Market Report. September 2010, Paris: IEA/OECD.

—— (2011) Key World Energy Statistics 2010. IEA/OECD, Paris (Also earlier editions).

IPCC (Intergovernmental Panel on Climate Change) (2007) *Fourth Assessment Report, Summary for Policy Makers*, Cambridge UK and New York: Cambridge University Press.

Irazabal, C. (2005) *City Making and Urban Governance in the Americas*, Aldershot, UK: Ashgate.

Ironmonger, D. and Norman, P. (2007) 'Travel behaviour of women, men and children: what changes and what remains the same?', paper presented at the 29th Annual Conference on Time Use Research, Washington DC, 17–19 October 2007.

Jacobs, M. (1991) *The Green Economy, Environment, Sustainable development and the politics of the future*, London: Pluto Press.

Jernelov, A. (2010) 'The threats from oil spills: now, then and in the future', *Ambio*, 39: 353–366.

Jessop, B. (1990) 'Regulation theories in retrospect and prospect', *Economy and Society*, 19: 153–209.

Jochem E. and Madlener, R. (2003) *The Forgotten Benefits of Climate Change Mitigation: Innovation, technological leapfrogging, employment and sustainable development*, Paris: OECD. http://www.oecd.org/dataoecd/6/49/19524534.pdf (accessed 18 April 2012).

de Jong, M., Rui, M., Stead, D., Yongchi, M. and Bao, X. (2010) 'Introducing public–private partnerships for metropolitan subways in China: What is the evidence?', *Journal of Transport Geography*, 18: 301–313.

Judt, T. (2011) *The Memory Chalet*, London: Vintage Books.

Kahn Ribeiro, S., Kobayashi, S., Beuthe, M. et al. (2007) 'Transport and its infrastructure', in B. Metz, O.R. Davidson, P.R. Bosch et al. (eds) *Climate Change 2007: Mitigation*. Cambridge, UK: Cambridge University Press, pp. 323–385.

Kant, I. (2002 [1798]) *Groundwork for the Metaphysics of Morals*, edited and translated by Allen W. Wood, with essays by J.B. Schneewind et al., New Haven: Yale University Press.

Karan, P.P. and Stapleton, K. (eds) (1997) *The Japanese City*, Lexington: The University Press of Kentucky.

Kaufmann, V. (2005) *Re-Thinking Mobility*, Aldershot UK: Ashgate.

Kearns, R., Collins, D. and Neuwelt, P. (2003) 'The walking school bus: extending children's geographies?', *Area*, 35: 285–292.

Kenworthy, J. and Laube, F. (2001) *The Millennium Cities Database for Sustainable Transport*, Brussels: Union Internationale de Transport Publique (UITP).

Kingham, S. and Ussher, S. (2007) 'An assessment of the benefits of the walking school bus in Christchurch, New Zealand', *Transportation Research Part A*, 41: 502–510.

Kitahara, T. (2003) 'Conserving walkable environments in Japan', in R. Tolley (ed.) *Sustainable Transport, Planning for walking and cycling in urban environments*, Boston: CRC Press and Cambridge UK: Woodhead Publishing, Ch 37, pp. 501–510.

Klare, M.T. (2008) *Rising Powers, Shrinking Planet: The new geopolitics of energy*, New York: Henry Holt.

Kloosterman, R.C. and Musterd, S. (2001) 'The polycentric urban region: towards a research agenda', *Urban Studies*, 38: 623–633.

Knapp, W. and Schmitt, P. (2003) 'Restructuring competitive metropolitan regions in North -West Europe: on territory and governance', *European Journal of Spatial Development*, 6: 1–42. http://www.nordregio.se/Global/EJSD/Refereed%20articles/refereed6.pdf (accessed 5 March 2005).

Kopits, E. and Cropper, M. (2005) 'Traffic fatalities and economic growth', *Accident Analysis and Prevention*, 37:169–178.

Kotkin, J. (2005) *The New Suburbanism: A realists guide to the American future,* Los Angeles: The Planning Centre. http://www.southbaycities.org/files/12.05%20Kotkin%20New%20Suburbanism%20Report.pdf (accessed 12 October 2011).

Krauss, C. (2011) 'A world reshaped by energy finds', International Herald Tribune, 27 Oct, p. 1.

Krugman, P. (1991) 'History and industry location: the case of the manufacturing belt', *American Economic Review*, 81: 80–83.

Kudsen, B. (2011) *Applying the SWIM2 Integrated Model for Freight planning in Oregon.* 13th TRB Transportation Planning Applications Conference. 11 May 2011, Reno, Nevada. Presentation accessed on 12–1–2012 at http://trb-appcon.org/TRB2011presentations/Session8/02_2011PlanningAppConf_KnudsonDRAFT.ppt.

Kukla, A. (2000) *Social Constructivism and the Philosophy of Science*, New York and London: Routledge.

Kutzbach, M. (2010) 'Megacities and megatraffic', *Access*, 37: 31–35.

Lash, H. (1976) *Planning in a Human Way: Personal reflections on the regional planning experience in Greater Vancouver*, Ottawa: Macmillan.

Lazanas, K. and Stone, J. (2010) 'An investigation of institutional arrangements for design and delivery of multimodal public transport in Melbourne', paper delivered at the 33rd Australasian Transport Research Forum, Canberra.

Lee, S., Seo, J.G. and Webster, C. (2006) 'The decentralising metropolis: economic diversity and commuting in the US suburbs', *Urban Studies*, 43: 2525–2549.

Legacy, C. (2010a) Examining the Knowledge Interface: Metropolitan plan-making and deliberative engagement, unpublished PhD thesis, Faculty of Architecture, Building and Planning, University of Melbourne.

—— (2010b) 'Investigating the knowledge interface between stakeholder engagement and plan-making', *Environment and Planning A*, 42: 2705–2720.

—— (2012) 'Achieving legitimacy through deliberative plan-making processes – lessons for metropolitan strategic planning', *Planning Theory and Practice*, 13: 71–87.

Legacy, C., Curtis, C. and Sturup, S. (2012) 'Good governance of land use and transport integration: An examination of Melbourne and Perth', *Transport Policy*, 19: 8–16.

Legacy, C., Sturup, S., Curtis, C. and Wigan, M.R. (2009) 'Delivering infrastructure and land use-transport integration policy: examining good governance issues in Melbourne and Perth' paper delivered at the (Vol. CDRom, pp. 14), Auckland NZ: Australian Transport Research Forum.

Leo, C. (1994) 'The urban economy and the power of the local state: the politics of planning in Edmonton and Vancouver', in F. Frisken (ed.) *The Changing Canadian Metropolis: A public policy perspective, Vol.* 2, Berkeley, CA: Institute for Governmental Studies Press.

—— (1998) 'Regional growth management regime: the case of Portland, Oregon', *Journal of Urban Affairs*, 20: 363–394.

Le Page, M. (2011a) 'Will there be more hurricanes?', *New Scientist*, 22 October: 42–43.

—— (2011b) 'Climate known: sea level is going to rise many metres', *New Scientist*, 22 October: 41.

Levinson, D. and Kumar. A. (1994) 'The rational locator: why travel times have remained stable', *Journal of the American Planning Association*, 60: 319–332.

Lew, Y.D. and Ang, C.I. (2010) 'The LTA journey: 15 years on', *Journeys*, November: 49–60.

Lindblom, C.E. (1977) *Politics and Markets, The world's political-economic systems*, New York: Basic Books.

—— (1990) *Inquiry and Change, The troubled attempt to understand and shape society*, New Haven: Yale University Press.

Lindblom, C.E. and Cohen, D.K. (1979) *Usable Knowledge: Social science and social problem solving*, New Haven: Yale University Press.

Lipietz, A. (1984) 'Accumulation, crises et sorties de crise', Research Paper no. 8409, Paris: CEPREMAP.

Loader, C. and Stanley, J. (2009) 'Growing bus patronage and addressing transport disadvantage –The Melbourne experience', *Transport Policy*, 16: 106–114.

Longhurst, D. (2008) *48 Months, 48 Minutes; Building the Perth to Mandurah Railway*, Perth WA: Rawlhouse Publishing Pty Ltd.

Low, N.P. (1991) *Planning, Politics and the State, Political foundations of planning thought*, London: Unwin-Hyman.

—— (2002) 'Ecosocialisation and environmental planning, a Polanyian approach', *Environment and Planning A*, 34: 43–60.

—— (2003) 'Is urban transport sustainable' in N.P. Low and B.J. Gleeson (eds) *Making Urban Transport Sustainable*, Basingstoke, UK: Palgrave-Macmillan, Chapter 1, pp. 1–22.

—— (2005) 'The discourse network: a way of understanding policy formation, stability and change in the networked polity', in L. Albrechts and S. Mandelbaum (eds) *The Network Society: A new context for planning?*. London and New York: Routledge, pp. 45–56.

Low, N.P. and Astle, R. (2009) 'Path dependence in urban transport: An institutional analysis of urban passenger transport in Melbourne, Australia, 1956–2006', *Transport Policy*, 16(2): 47–58.

Low, N.P. and Gleeson, B.J. (1998) *Justice, Society and Nature, An exploration of political ecology*, London and New York: Routledge.

Low, N.P. and Gleeson, B.J. (eds) (2003) *Making Urban Transport Sustainable*, Basingstoke, UK: Palgrave-Macmillan.

Low, N.P. and Odgers, J. (2012) 'Rethinking the cost of traffic congestion, lessons from Melbourne's city link toll roads', *Urban Policy and Research*, 30: 189–205.

Low, N.P. and Russell, E.W. (2008) *Submission on the East-West Link Needs Assessment Report, Investing in Transport*, GAMUT Working Paper (available from GAMUT Centre, the University of Melbourne, PO Box 4191, Victoria, Australia, 3052).

Low, N.P., Gleeson, B.J., Green, R. and Radovic, D. (2005) *The Green City, Sustainable homes, sustainable suburbs*, London and New York: Routledge and Sydney: UNSW Press.

Low, N.P., Gleeson, B.J. and Rush, E. (2005) 'A multivalent conception of path dependence: the case of transport planning in metropolitan Melbourne, Australia', *Environmental Sciences*, 4: 391–408.

LTA (Land Transport Authority, Singapore) (2008) *Land Transport Masterplan*, Singapore: Land Transport Authority.

—— (2011) *Singapore Land Transport Statistics in Brief*, Singapore: LTA.

Lynas, M. (2008) *Six Degrees, Our future on a hotter planet*, Washington DC: National Geographic.

Lyons, G, Jain, J. and Holley, D. (2007) 'The use of travel time by rail passengers in Great Britain', *Transportation Research A*, 41: 107–120.

Macintosh, A., Adams, N., Whyte, A. and Johnston, J. (2007) 'ePetitioning in the Scottish Parliament', in H. Chen, L. Brandt, V. Gregg, T.R.S. Dawes, E. Hovy, A. Macintosh and C.A. Larson (eds) *Digital Government: eGovernment Research, Case Studies and Implementation*, London, UK: Springer, pp. 487–501.

Macintosh, A., Robson, E., Smith, E. and Whyte, A. (2003) 'Electronic Democracy and Young People', *Social Science Computer Review*, 21: 43–54.

McCarthy, T. (2007) *Auto Mania, Cars, consumers and the environment*, New Haven and London: Yale University Press.

McClintock, H. (ed.) (2002) *Planning for Cycling, Principles, practice and solutions for urban planners*, Boston: CRC Press and Cambridge, UK: Woodhead Publishing.

McGee, T.G. (1994) 'Labour force change and mobility in the extended metropolitan regions of Asia', in R. Fuchs, E. Brennan, J. Chamie, F-C. Lo and J. Uitto (eds) *Mega City Growth and the Future*, Tokyo: United Nations University Press, pp. 62–102.

McKinsey and Co. (2008) An Australian Cost Curve for Greenhouse Gas Reduction, Sydney: McKinsey and Co. http://www.greenfleet.com.au/library/scripts/objectifyMedia.aspx?file=pdf/3/50.pdf&siteID=1&str_title=080215+McKinsey+Report+-+greenhouse.pdf (accessed 18 April 2012).

McMillan, T. (2005) 'Urban Form and a child's trip to school: the current literature and a framework for future research', *Journal of Planning Literature*, 19: 440–456.

McMillen, D. and McDonald, J. (1998) 'Suburban subcenters and employment density in metropolitan Chicago,' *Journal of Urban Economics*, 43: 157–180.

McVeigh, T. (2010) 'Cycling to School: is it such a terrible risk?', *The Observer*, 11 July 2010 http://www.guardian.co.uk/lifeandstyle/2010/jul/11/cycling-school-child-safety-schonrock/print (accessed18 October 2011).

Machiavelli, N. (1979 [1532]) *The Prince*, Oxford: Oxford University Press.

Mackett, R. (2004) *Making Children's Lives More Active*, London: Centre for Transportation Studies, University College London.

Mackett, R., Lucas, L., Paskins, J. and Turbin, J. (2004) *Walking Buses in Hertfordshire: Results from a postal survey*, London: Centre for Transport Studies, University College London.

Malakoff, D. (2011) 'Landmark study reveals an oil quagmire', *Science*, 333: 89.

Malone, K. (2012) 'Benign neglect is good for kids', Melbourne: *The Age*, 29 Feb, p. 13.

Malone, K. (2007) 'The Bubble-wrap generation: children growing up in walled gardens', *Environmental Education Research*, 13: 513–527.

Malone, K. and Rudner, J. (2011) 'Global perspectives on children's independent mobility: A socio-cultural comparison and theoretical discussion on children's lives in four countries in Asia and Africa', *Global Studies of Childhood*, 1: 207–242.

March, J. and Olsen, J. (1989) *Rediscovering Institutions: The organizational basis of politics*, New York: The Free Press.

—— (1995) *Democratic Governance*, New York: The Free Press.

—— (1996) 'Institutional perspectives on political institutions', *Governance: An International Journal of Policy and Administration*, 9: 247–264.

Marchetti, C. (1994) 'Anthropological invariants in travel behaviour', *Technological Forecasting and Social Change*, 47: 75–88.

Martell, L. (1994) *Ecology and Society: An introduction*, Oxford: Blackwell Publishers.

Martin, S. and Carlson, S. (2005) 'Barriers to children walking to or from school: United States, 2004', *Journal of the American Medical Association*, 294: 2160–2161.

Mastercard Worldwide (2008) *MasterCard Index. Worldwide Centers of Commerce*, http://www.mastercard.com/us/company/en/insights/pdfs/2008/MCWW_WCoC-Report_2008.pdf (accessed 24 September 2009).

Mataichi, S. (2010) 'JR Fusaiyo Mondai no Seiji Kaiketsu ni Atatte', Shakaishugi [*Socialism*] March, 2010.

Matthews, H. (1995) 'Living on the edge: children as outsiders', *Tijdschrift voor Economische en Sociale Geografie*, 86: 456–466.

Mayor of London (2004a) *The London Plan*. Section 5 'The Sub Regions, Central Activities Zone and the Government Growth Areas', http://www.london.gov.uk/thelondonplan/docs/londonplan08_ch5.pdf (accessed 12 October 2010).

Mayor of London (2004b) *The London Plan*. Section 3 'Connecting London', http://www.london.gov.uk/thelondonplan/docs/londonplan08_ch3.pdf. (accessed 12 October 2010).

Mees, P. (2000) *A Very Public Solution, Transport in the dispersed city*, Melbourne: Melbourne University Press.

—— (2002) Public transport privatisation in Melbourne: 'Teething problems' or something more serious? *25th Australasian Transport Research Forum*.

—— (2003) 'Paterson's curse: the attempt to revive metropolitan planning in Melbourne', *Urban Policy and Research*, 21: 437–440.

—— (2005) 'Privatisation of rail and tram services in Melbourne: What went wrong?', *Transport Reviews*, 25: 443–449.

—— (2010) *Transport for Suburbia: Beyond the automobile age*, London: Earthscan.

Mees, P., O'Connell, G. and Stone, H. (2008) 'Travel to work in Australian capital cities', 1976–2006', *Urban Policy and Research*, 26: 363–378.

Mees, P., Stone, J., Imran, M. and Neilson, G. (2010) *Public Transport Network Planning: a guide to best practice in New Zealand cities, NZTA Research Report No. 396*, Wellington: New Zealand Transport Agency.

Metz, D. (2008) *The Limits to Travel, How far will you go?*, London: Earthscan.

Miller, R.G. (2011) 'Future oil supply: The changing stance of the International Energy Agency', *Energy Policy*, 39: 1569–1974.

Miller, R. and Lessard, D.R. (2000) *The Strategic Management of Large Engineering Projects: Shaping institutions, risks and governance*, Cambridge, Mass.: MIT Press.

Mizuoka, F. (ed.) (2002) *Geography of Economy and Society*, Yuhikaku (in Japanese).

Mizuoka, F. (2008) 'Subsumption of space into society and alternative spatial strategy', *Geographishe Revue*, 10: 7–19.

Monheim, R. (2003) 'Visions for city traffic and mobility', in R. Tolley (ed.) *Sustainable Transport, Planning for walking and cycling in urban environments*, Boston: CRC Press and Cambridge UK: Woodhead Publishing, Ch 7, pp. 84–96.

Moriarty, P. (2002a) 'Household travel time and money expenditures', *Road and Transport Research*, 11(4): 2–14.

—— (2002b) 'Environmental sustainability of large Australian cities', *Urban Policy and Research*, 20: 233–44.

—— (2010) 'World car travel under environmental and resource constraints', paper presented at GAMUT International Conference, 2–4 June, Melbourne.

Moriarty, P. and Honnery, D. (2005) 'Determinants of urban travel in Australia'. Proceedings of the 28th Australian Transport Research Forum, 28–30 September, Sydney.

—— (2007) 'Australian car travel: an uncertain future', proceedings of the 30th Australian Transport Research Forum, 25–27 September, Melbourne.

—— (2008a) 'Low mobility: the future of transport', *Futures*, 40: 865–872.

—— (2008b) 'The prospects for global green car mobility', *Journal of Cleaner Production*, 16: 1717–1726.

—— (2009) 'Hydrogen's role in an uncertain energy future', *International Journal of Hydrogen Energy*, 34: 31–39.

—— (2011) *Rise and Fall of the Carbon Civilisation*, London: Springer.

—— (2012) 'Reducing personal mobility for climate change mitigation', in W.-Y. Chen, J.M. Seiner, T. Suzuki and M. Lackner (eds-in-chief) *Handbook of Climate Change Mitigation*, New York, US: Springer Science+Business Media, Chapter 51, pp. 1945–1979.

Morichi, S. (2005) 'Long-term strategy for transport systems in Asian megacities', *Journal of the Eastern Asia Society for Transportation Studies*, 6: 1–22.

Moroney, A. (2009) 'Healthy Spaces and Places: a national guide to designing places for active living', *Australian Planner*, 46: 12–15.

Morris, W. (2008) 'An Overview of the emerging Melbourne Alliance for Sustainable Transport and Urbanism' (A Better Plan for Melbourne), in GAMUT/VEIL (ed.) *Low Carbon Transport for Our Cities*, Melbourne.

Mosey, C. (2000) *Car Wars, Battles on the road to nowhere*, London: Vision.

Muller, H. (2004) 'Arguing, bargaining and all that: communicative action, rationalist theory and the logic of appropriateness in international relations', *European Journal of International Relations*, 10: 395–435.

Nabatchi, T. (2010) 'Deliberative democracy and citizenship: in search of the efficacy effect', *Journal of Public Deliberation*, 6, No.2 Article 8, http://services.bepress.com/jpd/vol6/iss2/art8 (accessed 19 April 2012).

Napoli, P.M. and Seaton, M. (2006) 'Necessary knowledge for communications policy: information inequalities and commercial data access and usage in the policymaking process', *Federal Communications Law Journal*, 59: 295–330.

National Heart Foundation of Australia (Victorian Division) (2009) *Modernising Victoria's Planning Act: submission to the Victorian Department of Planning and Community Development*. Available http://www.heartfoundation.org.au/SiteCollectionDocuments/Heart-Foundation-Planning-Act-Submission.pdf (accessed 6 April 2012).

NEA (National Environment Agency) (2010) *Singapore's Second National Communication under the United Nations Framework Convention on Climate Change*, Singapore.

Newman, P.W.G. and Kenworthy, J.R. (1989) *Cities and Automobile Dependence: A sourcebook*, Aldershot, UK: Gower Technical.

—— (2006) 'Urban design to reduce automobile dependence', *Opolis*, 2: 35–52.

—— (1999) *Sustainability and Cities: Overcoming automobile dependence*, Washington DC: Island Press.

Nielsen, G., Nelson, J.D., Mulley, C., Tegner, G., Lind, G. and Lange, T. (2005) *HiTrans Best Practice Guide No. 2, Public Transport: Planning the Networks*, Stavanger, Norway: European Union, Interreg IIIB.

North, D. (1990) *Institutions, Institutional Change and Economic Performance*, Cambridge UK: Cambridge University Press.

Notteboom, T.E. and Rodrigue, J.P. (2005) 'Port regionalization: towards a new phase in port development', *Maritime Policy and Management*, 32: 297–313.

O'Connor, J. (1984) *Accumulation Crisis*, Oxford: Basil Blackwell.

O'Connor, K. (2010) 'Global city regions and the location of logistics activity', *Journal of Transport Geography*, 18: 354–362.

O'Connor, K. and Healy E. (2004) 'Suburban development in Australia: A Melbourne case study', *Journal of European Planning Studies*, 12: 27–40.

O'Connor, K., Stimson, R. and Daly, M. (2001) *Australia's Changing Economic Geography. A society dividing*, Melbourne: Oxford University Press.

OECD (1995) *Urban Travel and Sustainable Development*, Paris: OECD Publications.

—— (2002) *Policy Instruments for Achieving Environmentally Sustainable Transport*, Paris: OECD Publications.

OECD Expert Group on Government Relations with Citizens and Civil Society (2003) *The Promise and Problems of E-democracy*, Paris: OECD Publications.

Olszewski, P.S. (2007) 'Singapore motorisation restraint and its implications on travel behaviour and urban sustainability', *Transportation*, 34: 319–335.

OPEC (Organization of the Petroleum Exporting Countries) (2010) *World Oil Outlook 2009*, Vienna, Austria: OPEC Secretariat.

Oreskes, N. and Conway, E.M. (2010) *Merchants of Doubt, How a handful of scientists obscured the truth on issues from tobacco to global warming*, New York, Berlin and London: Bloomsbury Press.

Pachauri, R.K. and Reisinger, A. (eds) (2007) *Synthesis Report, IPCC Fourth Assessment Report*, Geneva: Intergovernmental Panel on Climate Change.

Packer, J. (2008) *Mobility without Mayhem, Safety, cars and citizenship*, Durham and London: Duke University Press.

PACTS (Parliamentary Advisory Council on Transport Safety, UK) (1996) *Taking Action on Speeding*, London: UK Parliament.

Pan, H. (2011) *Implementing Sustainable Urban Travel Policies in China*, International Transport Forum, Discussion Paper No. 2011–12.

Pan, H. and Ming, Z. (2008) 'Rail transit impacts on land use, evidence from Shanghai, China', *Journal of the Transportation Research Board No. 2048*, Washington DC: Transportation Research Board of the National Academies.

Pan, H., Qing, S. and Ming Z. (2009) 'Influence of urban form on travel behaviour in four neighbourhoods of Shanghai', *Urban Studies*, 46: 275–294.

Parr, J.B. (2008) 'Cities and regions: problems and potentials', *Environment and Planning A*, 40: 3009–3026.

Peds Blog (2009) 'Blame the road: not the victim', January 20, 2009. Available http://peds.org/2009/01/ (accessed 18 October 2011).

Petzet, A. (2000) 'World resources estimate shows more liquids, slightly less gas', *Oil and Gas Journal*, 98: 79–80.

Phelps, N.A. and Parsons, N. (2003) 'Edge urban geographies: notes from the margins of Europe's capital cities', *Urban Studies*, 40: 1725–1749.

Pierson, P. (2000) 'Increasing returns, path dependence and the study of politics', *American Political Science Review*, 94: 251–267.

—— (2004) *Politics in Time, History, institutions and social analysis*, Princeton and Oxford: Princeton University Press.

Pisarski, A.E. (2002) Testimony on 'Mobility, Congestion and Intermodalism', United States Senate Committee on Environment and Public Works, March 19, 2002, Washington DC.

—— (2006) *Commuting in America III: The Third National Report on Commuting Patterns and Trends. Transit Cooperative Research Program (TCRP) Report 110/NCHRP Report 550*, published by the Transportation Research Board, Washington, http://onlinepubs.trb.org/onlinepubs/nchrp/CIAIIITopTenPresentation.pdf (accessed 17 July 2010).

Polanyi, K. (1957 [1944]) *The Great Transformation: The political and economic origins of our time*, Boston MA: Beacon Press.

Polanyi, M. (1976) *The Tacit Dimension*, London: Routledge and Kegan Paul.

Pooley, C., Turnbull, J. and Adams, M. (2006) 'The impact of new transport technologies on intraurban mobility: a view from the past', *Environment and Planning A*, 38: 253–267.

Porta, S., Crucitti, P. and Latora, V. (2006a) 'The network analysis of urban streets: a dual approach', *Physica A, Statistical Mechanics and its Applications, 369*: 853–866.

—— (2006b) The network analysis of urban streets: a primal approach, *Environment and Planning B: Planning and Design*, 33: 705–725.

Preston, J. and Rajé, F. (2007) 'Accessibility, mobility and transport-related social exclusion', *Journal of Transport Geography*, 15: 151–160.

Prezza, M., Alparone, F., Cristallo, C. and Luigi, S. (2005) 'Parental perception of social risk and of positive potentiality of outdoor autonomy for children: the development of two instruments', *Journal of Environmental Psychology*, 25: 437–453.

Pucher, J. and Kurth, S. (1996) 'Verkehrsverbund: the success of regional public transport in Germany, Austria and Switzerland', *Transport Policy*, 2: 279–291.

Punter, J. (2003) *The Vancouver Achievement: Urban planning and design*, Vancouver: University of British Columbia Press.

Putnam, H. (2004) *Ethics Without Ontology*, Cambridge: Harvard University Press.

Quiggin, J. (2010) *Zombie Economics, How Dead Ideas Still Walk Among Us*, Princeton NJ and Oxford: Princeton University Press.

Raab, C.D. (1992) 'Taking networks seriously: education policy in Britain', *European Journal of Political Research*, 21: 69–90.

Rawls, J. (2001) *Justice as Fairness*, Cambridge MA: Harvard University Press.

Reed, E.S. (1993) 'The intention to use a specific affordance: A conceptual framework for psychology', in R.H. Wozniak and K.W. Fischer (eds) *Development in Context: Acting and thinking in specific environments*, New Jersey: Lawrence Erlbaum Associates Inc, pp. 45–76.

Rees, W.E. (2003) 'Ecological footprints and urban transportation', in R. Tolley (ed.) *Sustainable Transport, Planning for walking and cycling in urban environments*, Boston: CRC Press and Cambridge UK: Woodhead Publishing, Chapter 1 pp. 3–19.

—— (2009) 'The ecological crisis and self-delusion: implications for the building sector', *Building Research and Information*, 37: 300–311.

Rein, W. and Roggenkamp, M. (1995) 'Can trams carry cargo, new logistics for urban areas', *World Transport Policy and Practice*, 1: 32–36.

Rhodes, R.A.W. and Marsh, D. (1992) 'New directions in the study of policy networks', *European Journal of Political Research*, 21: 181–205.

Richardson, J.J. and Jordan, A.G. (1979) *Governing Under Pressure: British politics in a post-parliamentary democracy*, Oxford: Martin Robertson.

Richmond, J.E.D. (2008) Transporting Singapore – The air conditioned nation. *Transport Reviews*, 28(3): 357–390.

Rietveld, P. and Stough, R.R. (eds) (2007) *Institutions and Sustainable Transport, Regulatory reform in advanced economies*, Cheltenham, UK: Edward Elgar.

Rifkin, J. (2009) *The Empathic Civilization, The race to global consciousness in a world in crisis*, New York: Tarcher-Penguin.

Riguelle, F., Thomas, I. and Verhetsel, A. (2007) 'Measuring urban polycentrism: a European case study and its implications', *Journal of Economic Geography*, 7: 193–216.

Risse, T. (2000) '"Lets Argue!": Communicative action in world politics', *International Organization*, 54: 1–39.

Rodrigue, J-P., Comtois, C. and Slack, B. (2009) *The Geography of Transport Systems*, New York: Routledge.

Rogers, J.M. (1989) 'Surrendering the ideal of disinterestedness in the policy research process: a cautionary note', *Knowledge in Society*, 1: 6–20.

Rorty, R. (1989) *Contingency, Irony and Solidarity*, New York: Cambridge University Press.

Rosenthal, S. (2002) 'Habermas, Dewey and the democratic self', in M. Aboulafia, M. Bookman and C. Kemp (eds) *Habermas and Pragmatism*, London and New York: Routledge, pp. 210–222.

Rudner, J. (2011) Conceptions of Risk and Children's Independent Mobility (unpublished PhD thesis). Melbourne: the University of Melbourne, Vic. 3010, Australia.

Rudner, J. and Malone, K. (2011) 'Childhood in the suburbs and the Australian dream: How has it impacted children's independent mobility?', *Global Studies of Childhood*, 1: 205–223.

Russell, B. (2009) *The Infrastructure Issues of Victoria's Population Growth*, GAMUT Working Paper, Melbourne: Australasian Centre for the Governance and Management of Urban Transport, University of Melbourne.

RWI (German Institute of Urban Affairs and NEA Transport Research and Training) (2010) Second State of European Cities Report. Research Project for the European Commission. DG Regional Policy. http://ec.europa.eu/regional_policy/sources/docgener/studies/pdf/urban/stateofcities_2010.pdf (accessed 2 October 2011).

Sabatier, P.A. (1991) 'Toward better theories of the policy process', *PS: Political Science & Politics*, 24: 147–156.

—— (1999) *Theories of the Policy Process*, Boulder, Colorado: Westview Press.

Sachs, W. (1992) *For Love of the Automobile, Looking back into the history of our desires*, (trans. D. Reneau), Berkeley CA: University of California Press.

—— (1999) *Planet Dialectics, Explorations in environment and development*, London and New York: Zed Books.

SACTRA (Standing Advisory Committee on Trunk Road Assessment) (1994) *Trunk Roads and the Generation of Traffic*, London: UK Department of Transport.

—— (1999) *Transport and the Economy*, London: UK Department of Transport.

Salet, W., Thornley, A. and Kruekels, A. (eds) (2003) *Metropolitan Governance and Spatial Planning*, London: Spon Press.

Sanders, L. (1997) 'Against deliberation', *Political Theory*, 25: 1–17.

Saward, M. (2000) 'Less than meets the eye: democratic legitimacy and deliberative theory', in M. Saward (ed.) *Democratic Innovation: Deliberation, representation and association*, London: Routledge, pp. 66–77.

Schaeffer, K.H. and Sclar, E. (1975) *Access for All: Transportation and urban growth*, Baltimore, USA: Penguin Books.

—— (2003) 'The Automobile Era: A cultural analysis', in A. Root (ed.) *Delivering Sustainable transport, A social science perspective*, Amsterdam: Pergamon, Elsevier Science, pp. 117–126.

Schäfer, A., Heywood, J.B., Jacoby, H.D. and Waitz, I.A. (2009) *Transport in a Climate-Constrained World*, Cambridge, MA. and London: MIT Press.

Scheurer, J. and Curtis, C. (2008) *Spatial Network Analysis of Multimodal Transport Systems: Developing a Strategic Planning Tool to Assess the Congruence of Movement and Urban Structure*. Research Monograph, Curtin University of Technology, Perth (WA), available online at www.abp.unimelb.edu.au/gamut/pdf/perth-snamuts-report.pdf.

Scheurer, J. and Porta, S. (2006) 'Centrality and Connectivity in Public Transport Networks and their Significance for Transport Sustainability in Cities'. *World Planning Schools Congress*, Mexico DF, 13–16 July.

Scheurer, J., Bergmaier, R. and McPherson, J. (2006) 'Keeping People Moving in Melbourne's North-East', Metropolitan Transport Forum (MTF), Melbourne (VIC).

Scheurer, J., Curtis, C. and Porta, S. (2007) 'Spatial network analysis of public transport systems: developing a strategic planning tool to assess the congruence of movement and urban structure in Australian cities', paper presented to the 30th Australasian Transport Research Forum and published in the proceedings, Melbourne, 25–27 September.

Schiller, L.P., Bruun, E.C. and Kenworthy, J.R. (2010) *An Introduction to Sustainable Transportation, Policy, planning and implementation*, London and Washington DC: Earthscan.

Scott, A.J. (1988) *Metropolis: From the division of labor to urban form*, Berkeley CA: University of California Press.

Scott, A.J. (ed.) (2001) *Global City-Region, Trends, theory, policy*, Oxford: Oxford University Press.

Searchinger, T., Heimlich, R., Houghton, R.A. et al. (2008) 'Use of U.S. croplands for biofuels increases greenhouse gases through emissions from land-use change', *Science*, 319: 1238–1240.

Self, P. (2000) *Rolling Back the Market, Economic dogma and political choice*, Basingstoke UK: Macmillan.

Selman, P. (1999) 'Three Decades of Environmental Planning: What Have We Really Learned?', in M. Kenny and J. Meadowcroft (eds) *Planning Sustainability*, London and New York: Routledge, pp. 148–174.

Sharp, L. (2005) *The Journey – Singapore's Land Transportation Story*, Belmont, Western Australia: International Publishing Pty Ltd.

Sheller, M. and Urry, J. (2000) 'The city and the car', *International Journal of Urban and Regional Research*, 24: 737–757.

—— (2006) 'The new mobilities paradigm', *Environment and Planning A*, 38: 207–226.

Short, J. and Murray, S. (2011) 'A frozen hell', *Nature*, 472: 162–163.

Simon, D. (1996) *Transport and Development in the Third World*, London and New York: Routledge.

Simpson, B. (1997) 'Towards the Participation of Children and Young People in Urban Planning and Design', *Urban Studies*, 34: 907–925.

Singapore, Government of (2008) *National Climate Change Strategy*, Singapore Government.

—— (2010) *Singapore's Second national Communication Under the United Nations Framework Convention on Climate Change*, Singapore: National Environment Agency.

Slack, B. (1999) 'Satellite terminals: a local solution to hub congestion', *Journal of Transport Geography*, 7: 241–246.

Sloman, L. (2003) 'The politics of changing to green modes', in R.S. Tolley (ed.) *Sustainable Transport: Planning for Walking and Cycling in Urban Environments*, Cambridge: Woodhead Publishing Limited.

Smith, D.A. and Timberlake, M. (2002) 'Hierarchies of dominance among world cities: a network approach', in S. Sassen (ed.) *Global Networks, Linked Cities*, London and New York: Routledge, pp. 117–141.

Smith, G. (2001) 'Taking deliberative seriously: institutional design and green politics', *Environmental Politics*, 10: 72–93.

—— (2003) *Deliberative Democracy and the Green Political Theory*, London and New York: Routledge.

Snyder, T. (2011) 'Georgia Mom Convicted of Vehicular Homicide for Crime of Crossing Street with Kids', *DC Streets Blog*, 14 July 2011. http://dc.streetsblog.org/2011/07/14/mother-convicted-of-vehicular-homicide-for-crossing-street-with-children/, (accessed 18 October 2011).

Solomon, S., Qin, D., Manning, M. et al. (2007) 'Technical Summary', in S. Solomon, D. Qin, M. Manning, et al. (eds) *Climate Change 2007: The physical science basis*, Cambridge, UK: Cambridge University Press.

Spence, N. and Frost, M. (1995) 'Work travel responses to changing workplaces and changing residences', in J. Brotchie, M. Batty, E. Blakely and P. Hall (eds) *Cities in Competition*, Melbourne: Longmans, pp. 359–381.

Sperling, D. and Gordon, D. (2008) 'Two billion cars: transforming a culture', *TR News*, Nov-Dec: 3–9.

Steffen, W., Sanderson, R.A., Tyson, P.D. and Jäger, J. (2005) *Global Change and the Earth System, A planet under pressure*, Berlin and New York: Springer.

Stern, N. (2006) *The Economics of Climate Change*, London: HM Treasury and Cabinet Office.

Stevenson, D. (1988) 'Civil Society and the Local State', in Rural Social and Economic Research Centre (ed.) *Agendas in Place: Urban and Cultural Planning for Cities and Regions*, Rockhampton: Central Queensland University Press.

Stokes, R. and Hill, R. (1992) 'The evolution of metropolitan planning in Western Australia', in D. Hedgcock and Y. Yftachel (eds) *Urban and Regional Planning in Western Australia: Historical and Critical Perspectives*, Perth WA: Paradigm Press.

Stone, C. (1989) *Regime Politics: Governing Atlanta, 1946–1988*, Lawrence: University of Kansas Press.

—— (1998) 'Regime analysis and the study of urban politics: a rejoinder', *Journal of Urban Affairs*, 20: 249–260.

Stone, J. (2008) Political factors in the rebuilding of mass transit: an investigation of failure in Melbourne since 1970 through comparisons with Perth and Vancouver, unpublished PhD thesis, Faculty of Life and Social Sciences, Swinburne University of Technology, Melbourne, (http://hdl.handle.net/1959.3/36049).

—— (2009a) 'Waning or just gone underground? Union power in public transport in Melbourne', *3rd State of Australian Cities National Conference*, Perth, University of Western Australia.

—— (2009b) 'Contrasts in reform: how the Cain and Burke years shaped public transport in Melbourne and Perth', *Urban Policy and Research*, 27: 419–434.

—— (2010) 'Turning over a new franchise: assessing the health of public transport management in Melbourne', paper presented at the 33rd Australian Transport Research Forum, Canberra: ATRF.

—— (2011) 'Can successful European models of public transport governance help to save Australian cities?', paper published in the Proceedings of the 5th State of Australian Cities Conference, November, Melbourne.

Stone, J. and Mees, P. (2011) 'Spatial distribution of the journey to work by sustainable modes in Australian cities' 34th Australasian Transport Research Forum, Adelaide, ATRF.

Stone, J., Mees, P. and Imran, I. (2012) 'Benchmarking the efficiency and effectiveness of public transport in New Zealand cities', *Urban Policy and Research*, 30: 207–222.

Strutynski, P. (1995) 'A new approach to reducing road freight transport', *World Transport Policy and Practice*, 1: 37–40.

Sturup, S. (2010) Managing the Mentalities of Megaprojects, unpublished PhD dissertation, Faculty of Architecture, Building and Planning, The University of Melbourne Vic. 3010, Australia.

Sudjic, D. (1992) *The 100 Mile City*, San Diego, CA: Harcourt Brace and Company.

Sundstrom, M. (2000) *A Brief Introduction: What is an epistemic community*, http://www.svet.lu.se/joluschema/epistcomm.pdf (accessed 1 February 2012).

Sunter, P. and Wigan, M.R. (2011) 'Enhancing community participation in metropolitan strategic transport planning through shared analysis, *Australian Transport Research*, 20: 24–35.

Takahashi, M. (2002) 'Kojin to Zentai no Niko Tairitsu wo Dou Koeruka', in K. Hirano (ed.) *Towards Reconstruction of the Economic Theory of Africa*, Institute of Development Economy (IDE-JETRO), pp. 113–146.

Thompson, G. and Matoff, T. (2003) 'Keeping up with the Joneses: radial vs. multidestinational transit in decentralizing regions', *Journal of the American Planning Association*, 69: 296–312.

Tickell, A. and Peck, J. (1995) 'Social Regulation after Fordism: regulation theory, neo-liberalism and global-local nexus', *Economy and Society*, 24: 357–386.

Tiebout, C. (1956) 'A Pure Theory of Local Expenditures', *Journal of Political Economy*, 64: 416–424.

Timperio, A., Crawford, D., Telford, A. and Salmon, J. (2004) 'Perceptions about the local neighbourhood and walking and cycling among children', *Preventive Medicine*, 38: 39–47.

Timperio, A., Salmon, J. and Ball, K. (2004) 'Evidence-based strategies to promote physical activity among children, adolescents and young adults: review and update', *Journal of Science and Medicine in Sport*, 7: 20–29 (supplement).

Tolley, R. (ed.) (2003) *Sustainable Transport, Planning for walking and cycling in urban environments*, Boston: CRC Press and Cambridge UK: Woodhead Publishing.

Torfing, J. (2001) 'Path-dependent Danish welfare reforms: the contribution of the new institutionalisms to understanding evolutionary change', *Scandinavian Political Studies*, 24: 277–309.

Torgerson, D. (1999) *The Promise of Green Politics: Environmentalism and the public sphere*, Durham USA: Duke University Press.

Transport Politic [The] (2010) 'Stations picked, huge automated transit project for Paris', www.thetransportpolitic.com/2010 (accessed 25 March 2010).

Tranter, P. (2004) *Effective Speeds: car costs are slowing us down*, Canberra: Australian Greenhouse Office.

Tranter, P. and Sharpe, S. (2008) 'Escaping Monstropolis: child-friendly cities, peak oil and Monsters Inc.', *Children's Geographies*, 6: 295–308.

Trapenberg Frick, K. (2008) 'The cost of the technological sublime: daring ingenuity and the new San Francisco-Oakland Bay Bridge', in H. Priemus, B. Flyvbjerg and B. van Wee (eds) *Decision-Making on Mega-Projects: Cost-benefit analysis, planning and innovation*, Cheltenham: Edward Elgar Publishing Limited, (pp. 239 – 262).

Troy, P.N. (1996) *The Perils of Urban Consolidation*, Annandale NSW, Australia: The Federation Press.

Turcotte, M. (2005) *The Time it Takes to Get to Work and Back*, Ottawa: Statistics Canada.

—— (2011) *Commuting to Work: Results of the 2010 General Social Survey*, Ottawa: Statistics Canada.

UNEP (United Nations Environment Programme) (1972) *Report of the UN Conference on the Human Environment*, www.unep.org/Documents.Multilingual/Default.Print.asp?documentid=97 (accessed 25 January 2012).

—— (2011) *Environment Assessment of Ogoniland*, Nairobi, Kenya: UNEP.

UNICEF [United Nations Children's Fund] (2008) *Child Friendly Cities*, Available http://www.childfriendlycities.org/ accessed 16 October 2011.

United Nations (1948) *Universal Declaration of Human Rights*, www.un.org/en/documents/udhr (accessed 25 January 2012).

—— (2009) *Copenhagen Accord*. http://unfccc.int/resource/docs/2009/cop15/eng/107.pdf (accessed 21 January 2010).

United States Census Bureau (2010) *Statistical abstract of the United States: 2010*, 129th edn. Washington DC: US Census Bureau.

United States Government, Federal Highway Administration (FHWA) (2009) *Southern California Regional Freight Study*, Office of Freight Management and Operations.

—— (2010) *FHWA Freight Planning. Discussion on Port-area Overweight Truck Corridor Programs*. Topic 3 Archives of FHWAFP@LISTSERV.UTK.EDU, August 2010; http://ops.fhwa.dot.gov/freight/freight_analysis/reg_ind_studies/so_cal_study.htm (accessed 17 April 2010).

Van den Bossche, P., Vergels, F., Van Mierlo, J., Matheys, J. and Van Autenboer, W. (2006) 'SUBAT: An assessment of sustainable battery technology', *Journal of Power Sources*, 162: 913–919.

Vega, A. and Reynolds-Feighan, A. (2008) 'Employment Sub-centres and travel-to-work mode choice in the Dublin Region', *Urban Studies*, 45: 1747–1768.

Verma, N. and Shin, H. (2005) 'Communicative action and the network society: a pragmatic marriage?', in L. Albrechts and S. Mandelbaum (eds) *The Network Society: A new context for planning?*, London and New York: Routledge, (pp. 9–23).

VicHealth (Victorian Health Promotion Foundation) (2008) *Streets Ahead: Supporting children to get active in their neighbourhoods*, Melbourne: Victorian Health Promotion Foundation.

Vigar, G. (2002) *The Politics of Mobility, Transport, the environment and public policy*, London and New York: Spon.

—— (2006) 'Deliberation, participation and learning in the development of regional strategies: transport policy making in North East England', *Planning Theory and Practice*, 7: 267–287.

Vuchic, V.R. (1999) *Transportation for Livable Cities*, New Brunswick, NJ: Rutgers University Press.

—— (2005) *Urban Transit: Operations, planning and economics*, Hoboken, New Jersey: John Wiley.

Wachs, M. and Crawford, M. (eds) (1992) *The Car and the City, The automobile, the built environment and daily urban life*, Ann Arbor: University of Michigan Press.

Wang, Y., Hayashi, Y., Kato, H. and Liu, C. (2011) 'Decomposition analysis of CO_2 emissions increase from the passenger transport sector in Shanghai, China', *International Journal of Urban Sciences*, 15: 121–136.

WAPC (Western Australian Planning Commission) (2004) *Network City: Community Planning Strategy for Perth and Peel*, Perth: Western Australian Planning Commission.

—— (2010) *Directions 2031 and Beyond*, Perth, Australia: Western Australian Planning Commission.

WAPC and DoP (Western Australian Planning Commission and Department of Planning) (2010) *Annual Report 2009 – 2010*, Perth: WAPC and DoP.

Ward, C. (1978) *The Child in the City*, London: The Architectural Press.

Ward, P.D. (2008) *Under a Green Sky, Global warming, the mass extinctions of the past and what they can tell us about our future*, New York: Collins and Smithsonian Books.

Warneryd, O. (1999) 'Transformation in the Swedish Urban System', in A.G. Aguilar and I. Escamilla (eds) *Problems of Megacities: Social Inequalities, Environmental Risk and Urban Governance*, Institute of Geography, Universidad Nacional Autonoma de Mexico, UNAM and Commission of Urban Development and Urban Life, IGU, pp. 577–587.

Warren, R., Price, J., Fischlin, A. et al. (2011) 'Increasing impacts of climate change upon ecosystems with increasing global mean temperature rise', *Climatic Change*, 106: 141–177.

Waste, R.J. (1987) *Power and Pluralism in American Cities: Researching the urban laboratory*, New York: Greenwood Press.

Weaver, P.M. (1998) 'The restructuring of transport, logistics, trade and industrial space use' in R.V. Ayres (ed.) *Eco-restructuring: implications for sustainable development*, Tokyo, New York and Paris: United Nations University Press.

Webster, D. (2002) *On the Edge: Shaping the future of peri-urban East Asia.* Urban Dynamics of East Asia Program. Asia Pacific Research Center. Stanford

University, http://aparc.stanford.edu/publications/on_the_edge_shaping_the_future_of_periurban_east_asia/ (accessed 20 November 2009).

Webster, D. and Muller, L. (2002) 'Challenges of Peri-Urbanisation in the Lower Yangtse Region: the Case of the Hangzhou-Ningbo Corridor'. Discussion Paper of the Asia Pacific Research Center, Stanford University. Stanford University, Stanford, Ca.

Welleman, T. (2002) 'An efficient means of transport: experiences with cycling policy in the Netherlands', in H. McClintock (ed.) *Planning for Cycling, Principles, practice and solutions for urban planners*, Boston: CRC Press and Cambridge, UK: Woodhead Publishing, pp. 192–208.

Whitelegg, J. (1997) *Critical Mass: Transport, environment and society in the 21st Century*, London: Pluto Press.

—— (2003) 'Transport in the European Union: time to decide' in N. Low and B. Gleeson (eds) *Making Urban Transport Sustainable*, Basingstoke UK: Palgrave Macmillan.

Whitzman, C. and Pike, L. (2007) 'From Battery-Reared to Free Range Children: Institutional Barriers and Enablers to Children's Independent Mobility in Victoria, Australia', Melbourne: Australasian Centre for Governance and Management of Urban Transportation (GAMUT Working Paper).

Whitzman, C., Worthington, M. and Mizrachi, D. (2009) *Walking the Walk: can Child-Friendly Cities promote Children's Independent Mobility?*, Melbourne: Australian Centre for the Governance and Management of Urban Transport (GAMUT Working Paper).

Wigan, M.R. (2008) *Response to the Victorian East-West Study by Professor Marcus Wigan*, GAMUT Working Paper. Melbourne Victoria: Victorian Department for Transport.

—— (2010) 'Identity, contestability and ethics of unified virtualisation of society', in K. Michael (ed.) *IEEE International Symposium on Technology and Society (ISTAS) 7–9 June 2010*, Wollongong, pp. 399–405.

—— (2011) 'The role of contestable processes in advancing sustainability in transport and planning', *Australian Transport Research*, 20: 40–48.

—— (2012) 'Uberveillance and faith-based organizations: a renewed moral imperative', in K. Michael and M.G. Michael (eds) *Uberveillance and the social implications of microchip implants: emerging technologies*, IGI Global.

Wilkinson, R. and Pickett, K. (2010) *The Spirit Level, Why equality is better for everyone*, London: Penguin Books.

Winchester, S. (1990) 'Leviathans of the Sky', *The Atlantic Monthly*, 266(4): 107–118.

Woodcock, J., Banister, D., Edwards, P., Prentice, A.M. and Roberts, I. (2007) 'Energy and transport', *The Lancet*, 370: 1078–1088.

Woodcock, J., Edwards P., Tonne, C., Armstrong, B.G., Ashiru, O., Banister, D., et al. (2009) 'Public health benefits of strategies to reduce greenhouse-gas emissions: urban land transport', *The Lancet*, 374: 1930–1943.

Woodlief, A. (1998) The path-dependent city', *Urban Affairs Review*, 33: 405–437.

WBCSD (World Business Council on Sustainable Development) (2004) *Mobility 2030: Meeting the challenges to sustainability (The Sustainable Mobility Project)*, WBCSD. http://www.wbcsd.org/Pages/EDocument/EDocumentDetails.aspx?ID=69&NoSearchContextKey=true (accessed 19 April 2012).

WCED (World Commission on Environment and Development) (1987) *Our Common Future*, Oxford and New York: Oxford University Press.

WTO (World Trade Organization) (2010) *World Trade Developments in 2009, International Trade Statistics 2010*, Table 1.4, www.wto.org/english/res_e/statis_e/its_world_trade_dev_e.htm (accessed 9 November, 2011).

Wu, F., Xu, J. and Gar-On, Y.A. (2007) *Urban Development in Post-Reform China, State, market and space*, London and New York: Routledge.

Wu, F. and Phelps, N.A. (2008) 'From suburbia to post-suburbia in China? Aspects of the transformation of the Beijing and Shanghai global city regions', *Built Environment*, 34, 464–481.

Wynne, B. (1991) 'Knowledges in context', *Science, Technology and Human Values*, 16: 111–121.

Yang, P.P. and Lew, S.H. (2009) 'An Asian Model of TOD: The Planning Integration of Singapore', in C. Curtis, J.L. Renne and L. Bertolini (eds) (2009) *Transit Oriented Development, Making it happen*, Farnham, UK and Burlington MA, USA: Ashgate, (pp. 91–108).

Yang, Q. (2008) 'How to improve the efficiency of railway freight transport in the urban area? Lessons from Los Angeles Alameda rail cargo expressway', *IEEE International Conference 2008*: 1246–1250.

Zahavi, Y. and Talvitie, A. (1980) 'Regularities in travel time and money expenditures', *Transportation Research Record*, 750: 13–19.

Zook, M.A. and Brunn, S.D. (2006) 'From podes to antipodes: positionalities and global airline geographies', *Annals of the Association of American Geographers*, 96: 471–490.

Index

For Product Safety Concerns and Information please contact our EU
representative GPSR@taylorandfrancis.com
Taylor & Francis Verlag GmbH, Kaufingerstraße 24, 80331 München, Germany

www.ingramcontent.com/pod-product-compliance
Lightning Source LLC
Chambersburg PA
CBHW071710180926
51764CB00004B/89
9780415529037